TABLE OF CONTENTS

PHOTO W. S. EVANS

BOG OR HEATH

Predominant woody plants represented are Pale Laurel, Bog Rosemary, Lamb-kill, Labrador Tea, Leather-leaf, Small Cranberry, and stunted specimens of Black Spruce and Tamarack.

DEDICATION

It is a pleasure to dedicate this publication to the *Josselyn Botanical Society* — a heterogeneous group of individuals representing many walks of life, bound together by a common bond, namely their great interest in plants. We owe much to this and similar botanical organizations for their dedicated effort in promoting the study of plants and teaching others, especially the younger generation whose interest is aroused and stimulated by the enthusiastic botanical pursuits of its members.

THE WOODY PLANTS OF SPHAGNOUS BOGS OF NORTHERN NEW ENGLAND AND ADJACENT CANADA[1]

Fully Illustrated

Fay Hyland[2]
Barbara Hoisington[3]
Illustrated by Laurel Smith[4]

INTRODUCTION

Historical

Bogs are fascinating places to visit! One may compare these sphagnum-covered areas with huge amphitheatres covered with wall-to-wall carpets intricately woven into multistructured mats. To the novice or one without botanical training, these areas might appear as monotonous assemblages of only a few species of stunted plants so similar in appearance as to appear homogeneous, but a keen observer will detect as many different species as might be found in a rich woods. By use of keys, descriptions, and illustrations provided, identification of all species is made easy and certain. Fifty or more different kinds of woody plants alone may be readily distinguished in bogs in the area covered by this publication. They are illustrated and described in this bulletin.

Sphagnum moss, orchids, cottongrass, and carnivorous plants (pitcher plants, sundews, bladderworts) are conspicuous nonwoody components of the mat, but by far the bulk of the tough fibrous material which binds the structure together is composed of the living and partially decayed roots and small stems of woody plants. Decay is slow due to the high acidity (pH4±) and low oxygen content of the substrate. It is questionable whether living roots are present in any but the superficial layers of the bog, the plants depending upon the relatively less acid rain water being held by the upper layers of sphagnum. Compared to well drained areas, the organic matter of bogs decomposes (mineralizes) extremely

---

[1] Supported in part by funds made available under the Hatch Act by the Life Sciences and Agriculture Experiment Station of the University of Maine at Orono.

[2] Professor Emeritus, Department of Botany and Plant Pathology, University of Maine at Orono.

[3] Botany student, graduate of the University of Vermont at Burlington.

[4] (Artist) Student of Botany and Art, University of Maine at Orono.

slowly due to the unfavorable conditions for microbial action. Available nitrogen and other nutrients are insufficient for vigorous growth of much of the bog flora, especially so of the larger woody plants which often develop into dwarfed and unhealthy individuals. Dead and dying trees of cedar, tamarack, and spruce are common. Much of the black spruce population results from rooting of the lower branches. Although cones with viable seeds are produced, the conditions necessary for germination and survival are often lacking in the sterile sphagnous mat which surrounds these trees. Most of the woody species included here may also be found in habitats where growing conditions are more favorable, but these particular plants have been singularly successful in adapting to the adverse conditions found in bogs. Several species which may not strictly be considered typical bog plants, but which usually are encountered along the wet margins of bogs, often invading the sphagnous mat, have been included in this bulletin. The bogs are continuously wet and support a luxurious surface growth of living sphagnum moss (*Sphagnum* spp.) which carpets the area and which, upon decay, adds considerably to the peaty material below the surface.

Based on radio carbon dating and other methods, these New England bogs are variously considered to be between 7,000 and 11,000 years old. They may exceed 30 feet in depth in some places. Studies of pollen grains of plants which grew in the area during development of the bogs reflects not only the types of plants present from time to time, but also the climates of the recent past.

The sphagnous areas of bogs often merge imperceptibly into swamps or open water on the side bordered by a low, flat terrain. In these areas, heath plants and several others are replaced by cattails, skunk cabbage, willows, grasses, sedges, rushes, and other typical aquatics, but the trees may persist. On the side of the bog bordered by a steep slope, the flora merges abruptly into the mesophytic type characteristic of the local area. Sphagnum is eliminated as light and moisture are reduced, and mineral soil becomes available to the plants. This bulletin is an illustrated guide to the identification of the woody flora commonly composing the fibrous mat of the typical continuously wet, acid, sphagnum-covered low moor type of bog, which is so common in the northeastern United States and adjacent Canada. High moor and alkaline bogs support a somewhat similar flora, but there are enough differences from the above to warrant a separate study.

Technical and Common Names

The technical names and order of arrangement of species in the text follow those of *Gray's Manual of Botany,* ed. 8, 1950, by M. L. Fernald. The sequence is essentially phylogenetic. By this method related groups are placed close together in families which, in turn, are arranged in order of their evolutionary advancement, beginning with the most primitive (or least specialized) and ending with the most advanced (or highly specialized). Most authors use this system of classification because of the advantage it offers over an alphabetical listing in the text. The common or vernacular names used in this bulletin were obtained from a variety of sources. Only one common name is used in conjunction with the technical name accompanying the plates and text, but several others are included at the end of each species description and all are listed in the index. Since many Canadian plants are featured in the bulletin, the common names used in Quebec and Newfoundland are designated as (Que.) and (Nfld.), respectively.

Illustrations

Nearly all the species figured in this bulletin were drawn from living material and depict the plant as it appeared at the time of collection. Occasionally, special diagnostic features are shown as insets. The illustrations should be used in portraying the over-all appearance of the plant, or portions of it, as it would appear in living condition in late spring or early summer when the specimen was collected. Many species are distinctive and can be readily distinguished by matching the specimen to be identified with the illustrations in the bulletin. However, developmental changes take place throughout the growing season, and, since only one stage of development is figured in the drawings, one should consider carefully the species descriptions accompanying the illustrations before reaching a final decision as to the identity of the specimen. Scales do not accompany the drawings. Actual measurements may be obtained from the text. So much variation exists in size, shape, pubescence, color, etc., in the different parts of plants during the growing season that a scale *per se* accompanying the illustrations is insufficient to depict the limits of variation encountered. For instance, there is a close resemblance among the species of blueberries (*Vaccinium* spp.) and the different species cannot therefore be positively determined by merely making comparisons with the illustrations. By carefully reading the accompanying species description, enough specific differences will be found to distinguish between them. The metric scale printed on the back cover will be useful in measuring parts of the plant.

How to Use the Key

With an unknown specimen before you, read both paragraphs numbered 1 in the key and decide which fits your plant. If it is the first number 1, then go on to the numbers 2 and again decide which is the correct choice. However, if the second number 1 fits your specimen better, then proceed with the number just below that (number 7). Keep on going that way, always choosing the number of each pair which fits your specimen, until you come to the name of the plant (printed in italics). Now turn to the page indicated and check your specimen with the description and illustration given there. Another (but less satisfactory) method is simply to leaf through the bulletin until you find a specimen which matches yours, then read the description on the facing page to see if it fits. A metric ruler is printed on the back page which may be used to measure parts of specimens.

KEY TO THE SPECIES

1. Seeds borne naked in woody or berry-like cones. Lvs. evergreen (except in Tamarack), awl-shaped, scale-like or linear. Trees, mostly resinous. Gymnosperms.
    2. Leaves needle-like, borne in fascicles of 2, bound together at the base by a membranaceous sheath; cones over 2.5 centimeters long. Jack Pine (*Pinus banksiana*) **(p. 6)**.
    2. Leaves linear, awl-shaped or scale-like, without a membranaceous sheath at base; cones less than 2.5 centimeters long.
        3. Lvs. linear, alternate.
            4. Lvs. deciduous, soft, flexible, borne in clusters of 8 to many on the spur branches, but scattered singly along the leading shoots. Cones erect. Tamarack (*Larix laricina*) **(p. 4)**.
            4. Lvs. evergreen, stiff, pointed, spirally arranged on all the branches. Cones pendent. Black Spruce (*Picea mariana*) **(p. 2)**.
        3. Lvs. scale-shaped or awl-shaped, opposite, or in threes.
            5. Branchlets rounded. Cone bluish or whitened, fleshy and berry-like. Northern Red Cedar (*Juniperus virginiana* var. *crebra*) **(p. 12)**.
            5. Branchlets flattened. Cone composed of leathery or slightly woody scales.
                6. Branchlets much flattened. Cones longer than broad, composed of flat uniformly thickened, tan or brown scales. Northern White-cedar (*Thuja occidentalis*) **(p. 8)**.
                6. Branchlets slightly flattened. Cones rounded, composed of bluish or whitened wedge-shaped or tack-shaped scales thickened at their outer ends and narrow toward the central point where they are attached. Atlantic White-cedar (*Chamaecyparis thyoides*) **(p. 10)**.
1. Seeds borne in an ovary. Lvs. deciduous (except those of several low shrubs), mostly broad and flat. Angiosperms.
    7. Tiny plant, parasitic on the branches of *Picea* and *Larix*. Dwarf Mistletoe (*Arceuthobium pusillum*) **(p. 22)**.
    7. Larger plants, not parasitic on the branches of conifers.
        8. Lvs. persistent, evergreen.
            9. Lvs. opposite or whorled.
                10. Lvs. either opposite or whorled, older leaves bending down on curved petioles. Twigs circular in cross section. Fruit axillary on slender, curved pedicels. Lambkill (*Kalmia angustifolia*) **(p. 60)**.

10. Lvs. all opposite, not bending down on curved petioles. Twigs 2-edged. Fruit terminal on long, slender, erect pedicels. Pale-Laurel (*Kalmia polifolia*) **(p. 62)**.

9. Lvs. alternate.
   11. Stems creeping or trailing, often rooting.
      12. Lvs. compound, stems and leaves with prickles. Swamp Dewberry (*Rubus hispidus*) **(p. 34)**.
      12. Lvs. simple, stems and leaves without prickles.
         13. Plants with wintergreen odor. Stems and leaves rough-hairy. Leaves not broader than long, fruit white. Creeping Snowberry (*Gaultheria hispidula*) **(p. 70)**.
         13. Plants without wintergreen odor. Stems and leaves without hairs. Leaves longer than broad, fruit red.
            14. Branchlets minutely glandular or pilose with viscid and sordid hairs. Leaves linear to narrowly elliptic, glabrous to glandular-pulverulent. Fruit a black or purple drupe. Black Crowberry (*Empetrum nigrum*) **(p. 38)**.
            14. Branchlets glabrous. Leaves ovate-oblong to ovate or triangular, or oblong-elliptic; glabrous, whitened or glaucous beneath. Fruit a red, acid berry.
               15. Lvs. oblong-elliptic, 6-17mm. long, 2-8mm. broad. Berry 10-20mm. thick, borne along side of stem on an elongate rachis 1-3cm. long. (Large Cranberry (*Vaccinium macrocarpon*) **(p. 88)**.
               15. Lvs. ovate-oblong to ovate or triangular, 3-8mm. long and 1-3mm. broad. Berry 5-8mm. in diameter, borne terminally on a rachis no more than 3 or 4mm. long. Small Cranberry (*Vaccinium oxycoccus*) **(p. 86)**.
   11. Stems erect.
      16. Lower surface of leaves covered with dense woolly, rusty-brown hairs, fragrant when crushed. Fruit a long slender capsule borne terminally, dehiscing from the base. Labrador Tea (*Ledum groenlandicum* **(p. 56)**.
      16. Lower surface of leaves without dense rusty hairs, not fragrant when crushed. Fruit neither a long slender capsule borne terminally nor dehiscing from the base.
         17. Lvs. thick, entire, and conspicuously whitened, especially below. Fruit a tomato-shaped capsule borne terminally on short, thick, recurved pedicels. Bog-Rosemary (*Andromeda glaucophylla)* **(p. 64.)**

17. Lvs. thinner, leathery, obscurely toothed, more or less scurfy, covered with rusty dots, becoming progressively smaller at tip of stem. Fruit pin cushion-shaped capsules, each with a pin-like style, borne on short straight pedicels. Leather-leaf (*Chamaedaphne calyculata*) **(p. 68)**.

8. Lvs. deciduous.
18. Lvs. compound.
19. Lvs. opposite. Black Ash (*Fraxinus nigra*) **(p. 90)**.
19. Lvs. alternate.
20. Stems and leaves with prickles. Northeastern Rose (*Rosa nitida*) **(p. 36)**.
20. Stems and leaves without prickles.
21. Lvs. pubescent with persistent thin, dry, sheathing stipules. Lfts. 5-7, margins revolute. Fruit an achene. Low much branched shrub. Shrubby Cinquefoil (*Potentilla fruticosa*) **(p. 32)**.
21. Lvs. glabrous, without persistent thin, dry, sheathing stipules. Lfts. 7-13, margins not revolute. Fruit a drupe. Small tree. POISONOUS TO TOUCH! Poison Sumac (*Rhus vernix*) **(p. 40)**.
18. Lvs. simple.
22. Lvs. lobed. Red Maple (*Acer rubrum*) **(p. 48)**.
22. Lvs. not lobed.
23. Lvs. opposite or whorled.
24. Lvs. mostly whorled.
25. Stems recurved, arching and rooting at the tips, the submersed ones spongy-thickened. Lvs. lanceolate. Fruit not in heads. Water-willow (*Decodon verticillatus*) **(p. 52)**.
25. Stems upright, neither arching nor rooting at the tips, not spongy-thickened. Lvs. ovate. Fruit in heads. Buttonbush (*Cephalanthus occidentalis*) **(p. 92)**.
24. Lvs. all opposite.
26. Lvs. entire and oblong. Fruit a berry. Brts. brittle and weak.
27. Lvs. tapering at base, glabrous or nearly so. Bark grayish. Berry red or purple. Swamp Fly-Honeysuckle (*Lonicera oblongifolia*) **(p. 96)**.
27. Lvs. rounded at base, hairy on both surfaces, ciliate. Bark or branchlets brown. Berry bluish-black. Mt. Fly-Honeysuckle (*Lonicera villosa*) **(p. 94)**.

26. Lvs. at least remotely serrate or dentate, elliptic-oblong, elliptic, or ovate. Fruit a drupe. Brts. stiff and strong.
    28. Lvs. elliptic-oblong, margin wavy or remotely toothed. Twigs terete. Witherod (*Viburnum cassinoides*) **(p. 98)**.
    28. Lvs. ovate and dentate. Twigs angled. Arrow-wood (*Viburnum recognitum*) **(p. 100)**.

23. Lvs. alternate.
  29. Lvs. entire.
    30. Pith of twigs diaphragmed. Large tree. Black Tupelo (*Nyssa sylvatica*) **(p. 54)**.
    30. Pith of twigs homogeneous. Shrub.
      31. Bud with a single cap-like scale. Bog Willow (*Salix pedicellaris* var. *hypoglauca*) **(p. 14)**.
      31. Buds with more than one bud scale.
        32. Lvs. mucronate; petioles and young twigs purple. Mt. Holly (*Nemopanthus mucronata*) **(p. 46)**.
        32. Lvs. not mucronate; petioles and young twigs not purple.
          33. Fruit a berry or berry-like drupe.
            34. Lvs. sprinkled with resinous dots (sticky when pinched between thumb and finger). Fruit a berry-like drupe containing ten seed-like nutlets.
              35. Inflorescence with conspicuous leaf-like bracts. Dwarf Huckleberry (*Gaylussacia dumosa*) **(p. 72)**.
              35. Inflorescence without conspicuous leaf-like bracts. Black Huckleberry (*Gaylussacia baccata*) **(p. 74)**.
            34. Lvs. not sprinkled with resinous dots (not sticky). Fruit a berry with more than ten tiny seeds.
              36. Shrub 1-4m. high. Lvs. downy or woolly underneath. Berries polished black, without bloom. Black Highbush Blueberry (*Vaccinium atrococcum*) **(p. 84)**.
              36. Shrub 2-6 dm. high. Lvs. downy on both sides. Berries, blue with much bloom (rarely black). Velvet-leaf Blueberry (*Vaccinium myrtilloides*) **(p. 76)**.
          33. Fruit a capsule.
            37. Capsule much longer than broad. Rhodora (*Rhododendron canadense*) **(p. 58)**.
            37. Capsule globular — Maleberry (*Lyonia ligustrina*) **(p. 66.)**.

29. Lvs. serrate or crenate.
  38. Bud with a single cap-like scale. Willow (*Salix spp.*).*
  38. Buds with more than one bud scale.
    39. Lvs. doubly serrate. Pith triangular in cross section. Speckled Alder (*Alnus rugosa*) **(p. 20)**.
    39. Lvs. not doubly serrate. Pith not triangular in cross section.
      40. Lvs. resin-dotted, aromatic, entire below the middle but with few coarse teeth near apex. Sweet Gale (*Myrica gale*) **(p. 16)**.
      40. Lvs. neither resin-dotted nor aromatic, serrate to the base.
        41. Lvs. with glands on upper surface of midrib of leaf near petiole.
          42. New branchlets, leaves, rachis, and pedicels more or less densely gray- or white-tomentose. Leaves varying from broadly oblanceolate to narrowly obovate or subelliptic, abruptly short acuminate at apex, tapering to the base. Fruit purple or purple-black. Purple Chokeberry (*Pyrus floribunda*) **(p. 28)**.
          42. New branchlets, leaves, rachis and pedicels essentially glabrous. Leaves variable but tending to be less obovate and abruptly pointed at tips — Fruit black. Black Chokeberry (*Pyrus melanocarpa*) **(p. 30)**.
        41. Lvs. without glands on upper surface of midrib of leaf near petiole.
          43. Lf. blades nearly as broad as long. Fruit a conelike aggregate of tiny nutlets or samaras. Bog Birch (*Betula pumila*) **(p. 18)**.
          43. Lvs. narrow, much longer than broad. Fruit not a conelike aggregate of tiny nutlets.
            44. Fruit a capsule. Maleberry (*Lyonia ligustrina*) **(p. 66)**.
            44. Fruit not a capsule.
              45. Fruit a berry or berry-like drupe.
                46. Fruit a berry-like drupe with few, rather large seeds.
                  47. Fruit black. Shrub with few branches, or not branched. Alder-leaved Buckthorn (*Rhamnus alnifolia*) **(p. 50)**.
                  47. Fruit red. Shrub much branched.
                    48. Lvs. thick, brittle, nearly glabrous above, pubescent, at least on the veins beneath. Fruit nearly sessile. Common and widespread. Common Winterberry (*Ilex verticillata*) **(p. 42)**.

---

*Although not strictly bog plants, several species of willow with serrate leaves are common in areas bordering the bog where there is enough mineral matter to support their growth. These species occasionally encroach upon the bog and become established where the sphagnous mat is sparse or lacking. A taxonomy manual should be consulted for species identification.

48. Lvs. thin, glabrous on both sides. Fruit not sessile; length of pedicels about equaling the diameter of the fruit. Local, restricted to few areas in southwestern Maine. Smooth Winterberry(*Ilex laevigata*) (**p. 44**).

46. Fruit a berry with many tiny seeds.
   49. Shrub less than 0.5 dm. high, much branched, freely stoloniferous.
      50. Lvs. green beneath. Fruit usually blue, with a bloom. Low Sweet Blueberry (*Vaccinium angustifolium* var. *laevifolium*) (**p. 78**).
      50. Lvs. glaucous beneath. Fruit usually black, without a bloom. Black-fruited Blueberry (*Vaccinium angustifolium* var. *nigrum*) (**p. 80**).
   49. Shrub to 4m. high forming compact or open clumps. Highbush Blueberry (*Vaccinium corymbosum* vars.) (**p. 82**).

45. Fruit a follicle.
   51. Lvs. glabrous, broadly oblanceolate or obovate. Twigs glabrous, tan-brown. Meadow-sweet (*Spiraea latifolia*) (**p. 24**).
   51. Lvs. tomentose, broad elliptic to elliptic-oblong. Twigs tomentose, purplish. Hardhack (*Spiraea tomentosa*) (**p. 26**).

DESCRIPTION OF SPECIES AND ILLUSTRATIONS

PINACEAE (Pine Family)
Black Spruce
Picea mariana (Mill.) BSP.

Habit.- Resinous evergreen conifer, up to 30 meters high (much dwarfed in bogs where its lower branches often root in the sphagnum and perpetuate the tree in this manner); a ragged-appearing tree, forming a narrow, often irregular head; twigs pubescent.

Leaves.- Evergreen, simple, bluish-green with whitish bloom, alternate, 6-13 millimeters long, needle-shaped, 4-sided, attached to the twigs by short, peg-like stalks which remain as rough projections when the leaves are ultimately shed (or specimens are dried).

Flowers.- Monoecious: male flowers catkin-like, dark red; female flowers cone-like, purplish, April-May.

Fruit.- Ovoid, dull grayish or purplish-brown, blunt-pointed, scaly cones, 2-3 centimeters long, borne on short, curved stalks, maturing in 1 season but persisting for several years; cone-scales stiff, with erose or dentate margins, narrowed toward the apex.

Habitat.- Adjusting to bogs, as does Tamarack, where it occurs singly or in small isolated groups over wide areas; trees often distorted and stunted when growing in deep sphagnum but attaining better size and form near the wet bog borders; elsewhere on cool slopes and in humus-filled depressions at the higher elevations.

Notes.-Resembles red spruce (*Picea rubens* Sarg.) and, at times, difficult to separate from it. Frequently hybridizing with red spruce, especially in the north.

Other Names.- Bog Spruce, Eastern Spruce, Épinette noire (Que.), Shortleaf Black Spruce, Swamp Spruce.

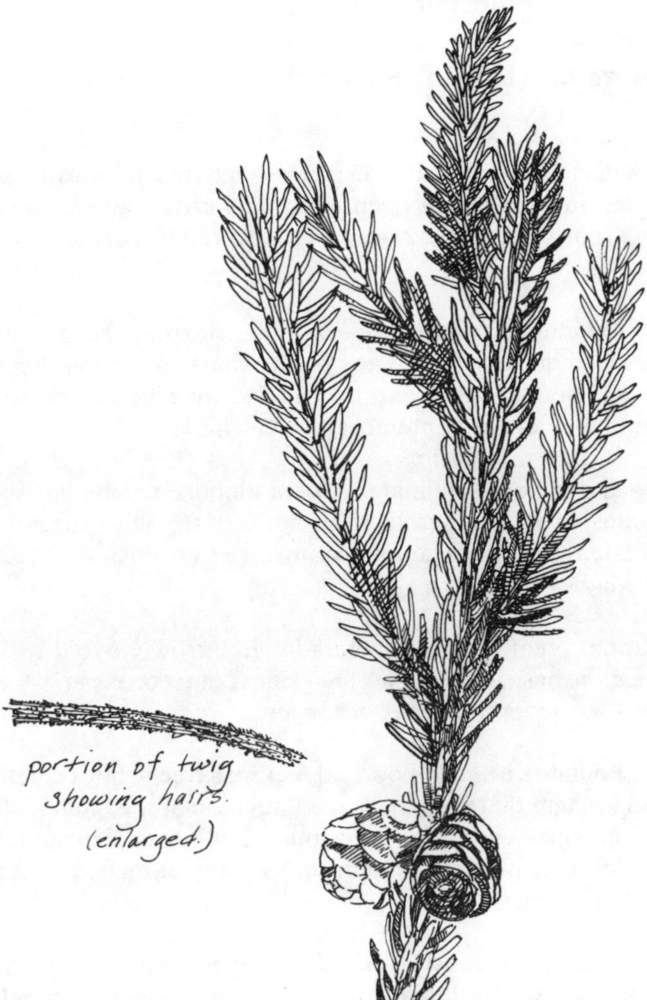
portion of twig
showing hairs.
(enlarged.)

PINACEAE (Pine Family)
Tamarack
Larix laricina (DuRoi) K. Koch

Habit.- A deciduous conifer to 35 meters high (much dwarfed, scraggly, and stunted in bogs), with open pyramidal crown and shallow, wide-spreading root system; bark scaly, gray to reddish brown; wood hard and resinous.

Leaves.- Deciduous, simple, linear, soft, flexible, 1-2.5 centimeters long, arranged in clusters of many on the short spurs, spirally arranged on the long shoots of the season; 3-angled, turning yellow and falling from the branchlets in September to November.

Flowers.- the yellow staminate flowers globose to oblong, stalked or sessile, consisting of numerous short-stalked, spirally arranged anthers; the pistillate flowers subglobose, crimson or greenish with red bracts, April to May.

Fruit.- Cones erect, 1-2 centimeters long, narrowly ovoid with visible bracts near the base, cone scales few, shiny, tan; cones persistent on the branchlets for several months or longer.

Habitat.- Found as a small, poorly-developed tree in bogs chiefly on the hummocks where there is a thick accumulation of sphagnum; also along the wet margins where the sphagnum blanket is thinner and some mineral soil is available; common in swamps elsewhere where it becomes a fairly large tree.

Notes.- Often (erroneously) called "juniper" (at least in Northern Maine), and known by Indians in New York State as "Ka-neh-tens". Formerly used as "ship knees" in wooden vessels.

Other Names.- Alaska Larch, American Larch, Black Larch, Eastern Larch, Épinette rouge (Que.), Hackmatack, "Juniper".

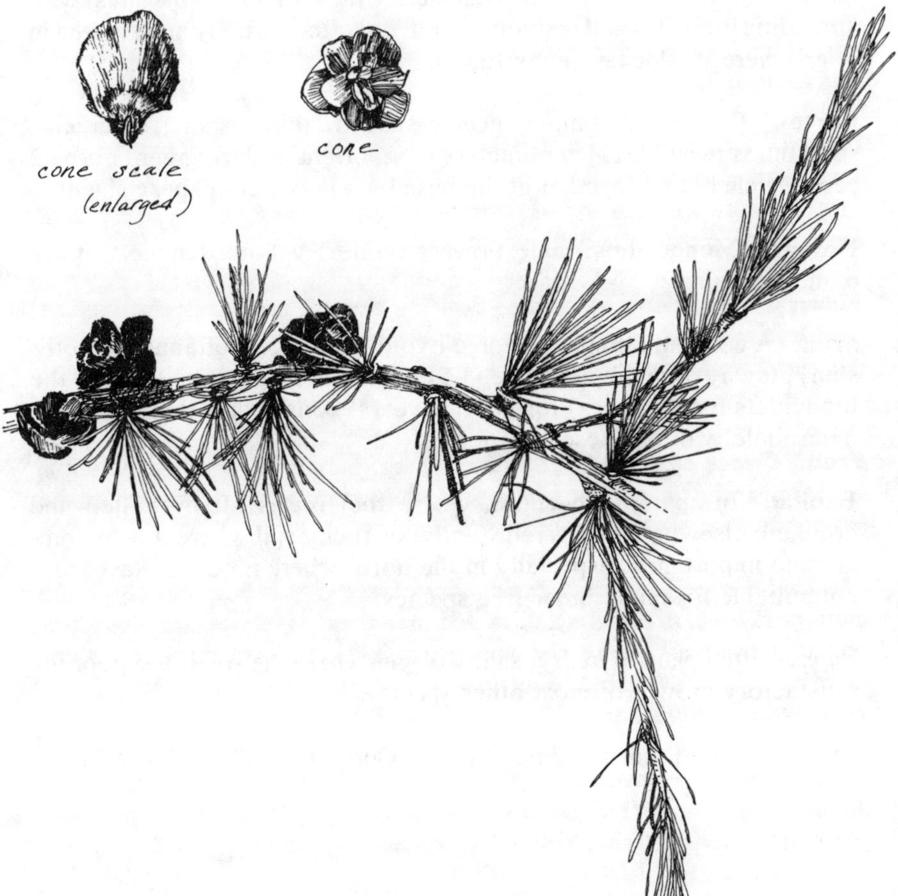
cone scale
(enlarged)
cone

PINACEAE (Pine Family)
Jack Pine
Pinus banksiana Lamb.

Habit.- An evergreen conifer to 25 meters high (on favorable sites) with spreading branches and resinous wood, but often shrubby and stunted in bogs where it is occasionally found.

Leaves.- Evergreen, simple, needle-shaped, thick, short, divergent, sometimes twisted, 2-4 centimeters long, bright or dark green, borne 2 per fascicle bound together at the base by a persistent papery sheath.

Flowers.- Monoecious: male flowers (cones) yellow; female flowers (conelets) purple, May-June.

Fruit.- A conical, oblong, or conic-ovoid, curved, serotinous, smooth, shiny, tawny or yellowish cone, 3-5 centimeters long, persisting on the branchlets (or branches) for several years; cone-scales thin, minutely mucronate with obsolescent prickles.

Habitat.- In sparse sphagnum, where the tree is often stunted and crooked; elsewhere on barren, sandy, or rocky soil where it is of commercial importance, especially in the north where it covers vast areas not suitable for many other tree species.

Notes.- Often planted in dry, sandy places where the soil is too poor for satisfactory growth of most other species.

Other Names.- Banksian Pine, Cyprès (Que.), Gray Pine, Scrub Pine.

PINACEAE (Pine Family)
Northern White-cedar
Thuja occidentalis L.

Habit.- Pyramidal, strong-scented, evergreen conifer to 20 meters high (smaller and stunted in bogs) with short, spreading branches, strongly flattened branchlets (all in one plane) clothed with small, scale-like, overlapping leaves; trunks buttressed, with shreddy, red-brown, soft bark.

Leaves.- Evergreen, simple, small, scale-like and opposite, overlapping in 4 rows, glandular on the back, completely clothing the twigs and young branchlets which are disposed in flat, horizontal sprays.

Flowers.- Monoecious, tiny, cone-like, composed of few scales: male flowers yellow; female flowers purplish or reddish, April or May.

Fruit.- Cones erect, oblong or elliptic-oblong, light or yellowish brown, 8 millimeters long, with 8-10 thin, flat scales (only 4 fertile), attached at different points to an elongate axis, maturing in one growing season and remaining on the branchlets for about a year.

Habitat.- Often getting a "foothold" here and there in the deep sphagnum where it is of poor, stunted growth; developing better near the border of the bog where mineral matter is available; attaining better development in swamps and on cool rocky banks, especially where the soil is limy.

Notes.- Wood scented, light, soft, and durable; often used for poles, posts, railroad ties, etc.; important to wildlife as a browse and for shelter.

Other Names.- American Arbor Vitae, Balai (Que.), Cèdre, Eastern Arbor Vitae, Eastern White-cedar, Oo-soo-ha-tah ("Feather-leaf"), Swamp-cedar.

PINACEAE (Pine Family)
Atlantic White-cedar
Chamaecyparis thyoides (L.) BSP.

Habit.- Strong-scented evergreen conifer to 25 meters high (often dwarfed in bogs), with upright, horizontal spreading, irregularly arranged branches forming a narrow spire-like head; bark of trunk shreddy, soft, reddish brown.

Leaves.- Evergreen, simple, minute; scale-like and opposite, completely clothing the twigs and young branchlets (which are slightly flattened but not disposed in horizontal sprays, as in *Thuja*).

Flowers.- Monoecious, tiny, terminal, cone-like: the male flowers, yellow, the female flowers, purplish brown, March or April.

Fruit.- Cones maturing in one growing season, subglobose, bluish purple, bloomy, 6-9 millimeters in diameter, not opening until maturity in the fall; cone-scales thick, peltate with terminal boss, wedge-shaped, attached at a central point (as opposed to the elongate cone-axis in *Thuja*). Occasional in sphagnum but more often in coastal swamps where it becomes commercially important southward.

Notes.- The scented wood is exceedingly durable; used for the same purposes as Northern White-cedar.

Other Names.- Coast White-cedar, False Cypress, Swamp-cedar, Southern White-cedar, White-cedar.

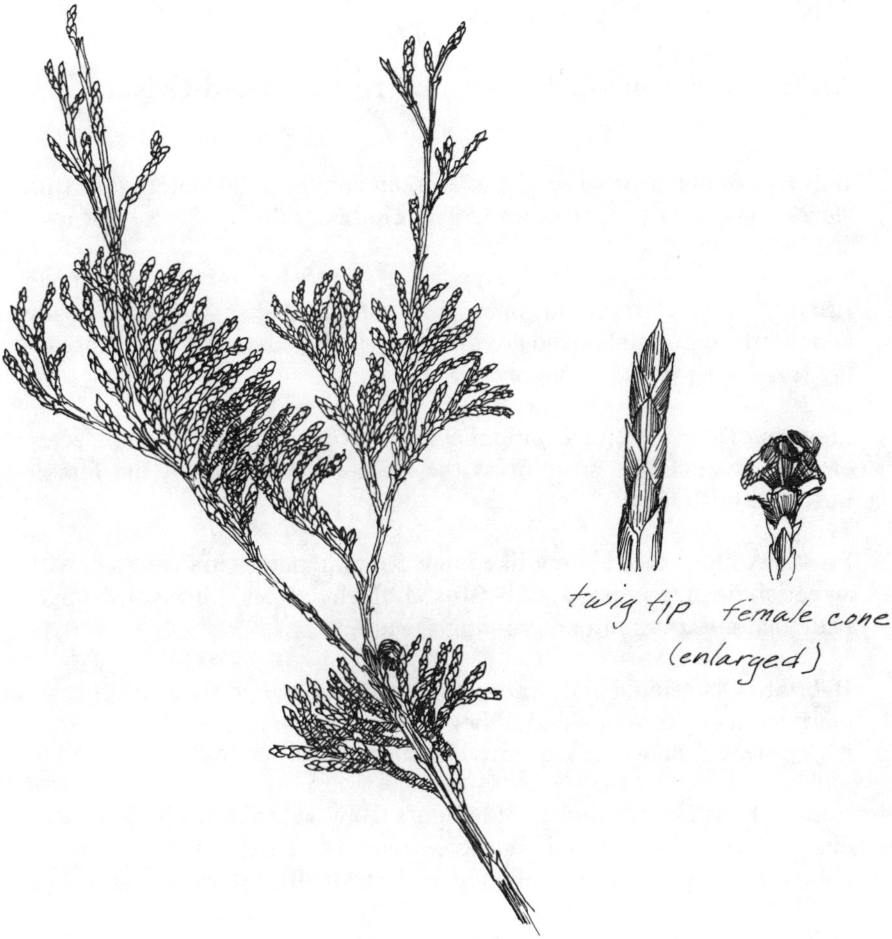
twig tip
female cone
(enlarged)

PINACEAE (Pine Family)
Northern Red Cedar
Juniperis virginiana L. var. *crebra* Fern. and Grisc.

Habit.- Columnar or spire-like evergreen conifer to 30 meters tall, with slender 4-angled or terete branchlets clothed with small scaly or awl-shaped leaves.

Leaves.- evergreen, simple, small, dimorphic, either scale-like, opposite and tightly appressed on old growth, or awl-shaped, ternate, and spreading (on young plants or vigorous shoots).

Flowers.- Dioecious (rarely monoecious), borne laterally in small, scaly catkins or cone-like structures: the male flowers yellow, the female flowers bluish, May.

Fruit.- A blue-black, berry-like cone 5-6 millimeters in diameter, with sweetish flesh, composed of 3-6 fused, fleshy scales, borne on straight pedicels, maturing in one growing season.

Habitat.- Occasional in the more open places in the bog or along the wet margins; more common and of better growth in dry, open woods, or on rocky slopes, and sandy or gravelly barrens, in neutral or acid soils.

Notes.- Thrives on a variety of habitats from wet to dry. The beautiful, characteristically scented, durable wood is valuable for chests and cabinets. Trees are often infected with unsightly cedar-rust galls.

Other Names.- Cèdre Rouge (Que.), Eastern Red-cedar, Red-cedar, Red Juniper, Savin.

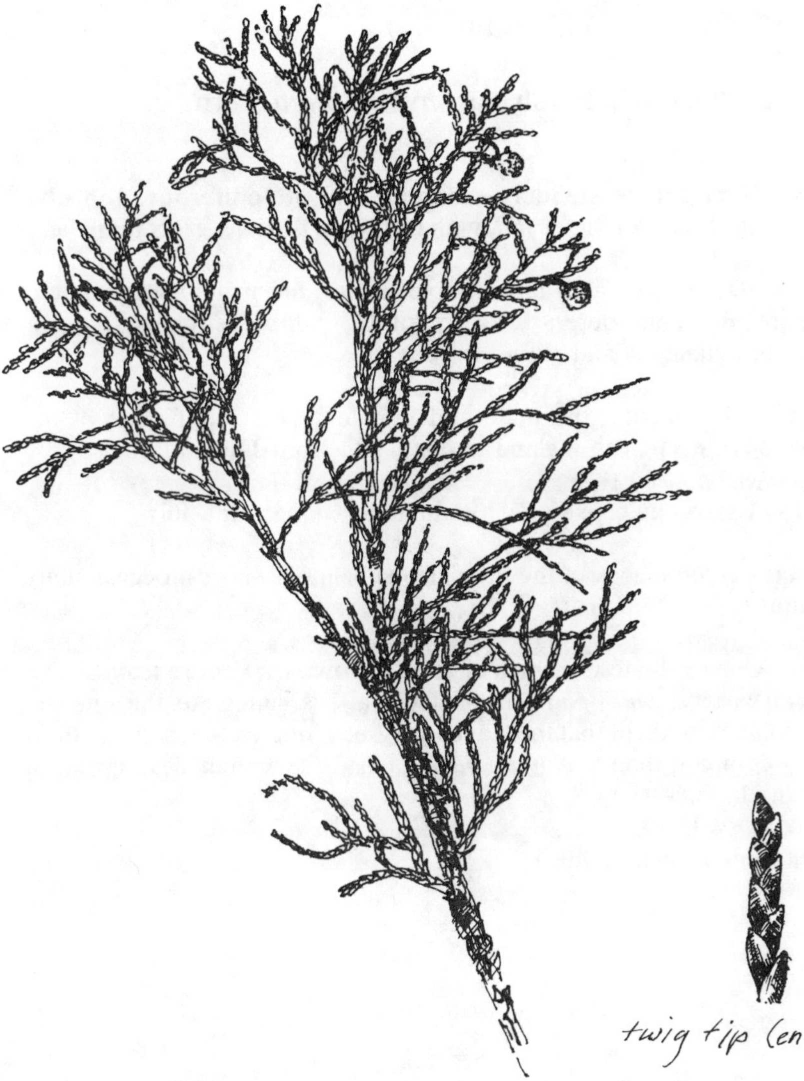

twig tip (enlarged)

SALICACEAE (Willow Family)
Bog Willow
Salix pedicellaris Pursh var. *hypoglauca* Fern.

Habit.- Upright or slender, creeping and stoloniferous, loosely-branching shrub to 1 meter tall, with glabrous, flexible, erect branches.

Leaves.- Deciduous, alternate, simple, entire, linear-oblong to elliptic-obovate, 1-5 centimeters long, revolute, without stipules, pale or whitened, glaucous and glabrous beneath.

Flowers.- In aments; the male and female sexes on different plants (dioecious), each with a gland at base, late April-July.

Fruit.- A glabrous capsule, 5-10 millimeters long, June-July.

Habitat.- Open places in the bog; also sphagnous shores, occasionally subalpine.

Notes.- One of the few species of local willows with entire leaves. The typical variety, *Salix pedicellaris* (typical), is similar to the one described above except that the leaves are green, not whitened, beneath. It is less common than var. *hypoglauca* Fern., hence our description of only the latter variety.

Other Name.- Saule (Que.).

Capsules
(enlarged)

MYRICACEAE (Sweet Gale Family)
Sweet Gale
Myrica gale L.

Habit.- Aromatic shrub to 2 meters high with brown, wand-like, strongly ascending branches.

Leaves.- Deciduous, alternate, simple, cuneate-oblanceolate, dark green and glabrous above, grayish-glabrous or more or less pubescent beneath, toothed toward the apex, 3-6 centimeters long.

Flowers.- Male catkins 7-10 millimeters long; female catkins cone-like, dense, 8-10 millimeters long, April-June.

Fruit.- Dense, cone-like, 8-10 millimeters long; individual fruits ovoid, imbricated, 2-winged nutlets, resin-dotted, borne in firm, cone-like structures at the ends of the branchlets.

Habitat.- Open places in the bog or along the wet margins, often growing in water; also on wet shores and in swamps.

Notes.- Scented shrub with pleasant odor. One of the most common woody plants on wet shores.

Other Names.- Bois-sent-bon (Que.), ''Meadow Fern'', Piment Royal (Que.).

BETULACEAE (Birch Family)
Bog Birch
Betula pumila L.

Habit.- Erect or prostrate and matted, 0.5-5 meters high, with glabrous to densely pubescent or tomentose, glandless branchlets.

Leaves.- Deciduous, alternate, simple, coarsely or dentate-serrate, suborbicular to broad-elliptic or obovate, reticulate-veined, 0.8-7 centimeters long.

Flowers.- Staminate catkins elongated, formed in autumn and remaining naked during the winter; pistillate catkins terminating short lateral branches (spur shoots) of the season, May-June.

Fruit.- A minute nutlet or samara, borne in sessile, erect, fragile strobiles 0.7-3 centimeters long and 5-9 millimeters thick, maturing in the fall.

Habitat.- Growing in the sphagnous areas; also found in wooded swamps, especially if there is available calcium.

Other Names.- Bouleau (Que.), Low Birch, Swamp Birch.

BETULACEAE (Birch Family)
Speckled Alder
Alnus rugosa (Du Roi) Spreng.

Habit.- Spreading or loosely ascending shrub to 8 meters high, the brown or blackish-gray bark marked with whitish linear lenticels 7 millimeters or more long; pith of twigs triangular in cross-section.

Leaves.- Deciduous, alternate, simple, oval, ovate, rounded to subcordate at bases, mostly doubly serrate or serrate-dentate, often with a wavy margin, scalariform-reticulate, rugose, 5-10 centimeters long.

Flowers.- In preformed catkins; the male flowers purplish, conspicuous, with 4 stamens; the female flowers smaller, developing into woody, cone-like structures in fruit, March-early June.

Fruit.- A woody strobile or cone-like structure about 1.5 centimeters long, borne in inflorescences of 4-8 on short-stalked or sessile pedicels, autumn.

Habitat.- A common shrub along the bog margins, often persisting in the sphagnous mat; larger and more abundant on low ground, in swamps, and along the margins of the water courses.

Notes.- Often forming dense thickets in wet meadows, etc. where few other plants except willows (*Salix* spp.) would thrive, thus forming favorable habitats for wildlife.

Other Names.- Aulne Blancharte (Que.), Hoary Alder, Verne (Que.).

twig cross-section
showing triangular
pith. (enlarged)

LORANTHACEAE (Mistletoe Family)
Dwarf Mistletoe
Arceuthobium pusillum Peck

Habit.- Tiny, fleshy to slightly woody plant, only 0.6-2 centimeters high, commonly parasitic on the branchlets of spruce (*Picea*) rarely on tamarack (*Larix*) and pine (*Pinus*); stems glabrous, olive to brownish or purplish, nearly terete or 4-angled in cross-section, simple or branched, brittle at base, arising from the cambium of the host and perenniating in scattered clusters.

Leaves.- Tiny, entire, leathery, appressed, scale-like, opposite, without stipules, connate, obtuse, suborbicular, olive or brown.

Flowers.- Dioecious, tiny, borne in the axils of the scales, April to June.

Fruit.- Ovoid or ellipsoid flattened berry or drupe, 2-3.5 millimeters long.

Habitat.- Parasitic on the branches of *Picea, Larix* and *Pinus* (in Europe, on *Juniperus*); occasional and sometimes abundant, especially on *Picea mariana* (Mill.) BSP.

Notes.- The only parasitic woody plant in our area; often causing "witches' brooms" on the coniferous host.

Other Names.- Petit Gui (Que.).

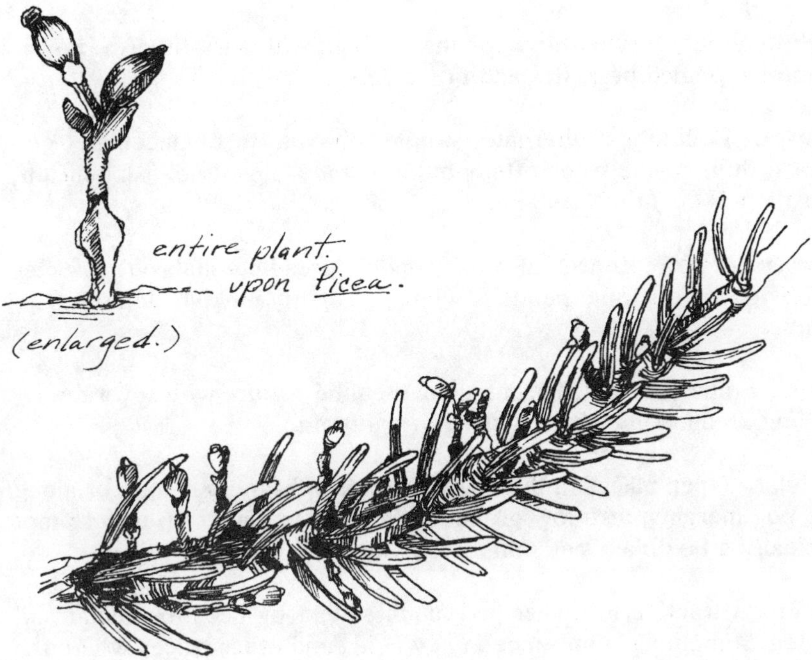

entire plant.
upon Picea.
(enlarged.)

ROSACEAE (Rose Family)
Meadow-sweet
Spiraea latifolia Ait. Borkh.

Habit.- Erect, wiry shrub to 1.5 meters high with dark reddish-brown, glabrous, angled branches and branchlets.

Leaves.- Deciduous, alternate, simple, obovate to oblanceolate, glabrous, thin, coarsely and often doubly serrate, slightly bluish beneath, mostly 1.5-7 centimeters long.

Flowers.- Inflorescences of broad, open, spreading, glabrous panicles 0.5-3 decimeters long; petals 5, white or slightly pinkish, June to September.

Fruit.- Follicles 5-8, dehiscent along the inner suture with several tiny, oblong seeds, somewhat persistent in autumn.

Habitat.- Open places in the bog where the sphagnum is thin, or along the bog margins; also low ground or dry, open places elsewhere; not typically a bog plant but sometimes found there.

Notes.- Attractive in flower and much sought by bees for nectar and pollen; sometimes a nuisance in hay fields and other places where the tough, wiry stems foul up the knives of the mowing machines.

Other Names.- Thé du Canada (Que.).

Fruit (follicle)
(enlarged)
flower
(enlarged.)

ROSACEAE (Rose Family).
Hardhack
Spiraea tomentosa L.

Habit.- Upright shrub to 1.5 meters tall with angled, dark purplish branchlets (but color often obscured by a coating of dense, matted brownish or tawny, woolly tomentum); branches tough and wiry.

Leaves.- Deciduous, alternate, simple, unequally and often doubly serrate, ovate to ovate-oblong, rugulose above, densely woolly with yellowish or grayish tomentose hairs beneath, 3-7 centimeters long.

Flowers.- Inflorescences of closed, compact, narrow, spire-like brownish, tomentose panicles, 8-20 centimeters long; petals 5, deep rose or rose-purple (rarely white), showy, attractive, July-September.

Fruit.- Follicles 5-8, dehiscent along the inner suture, with several, tiny, oblong seeds, somewhat persistent, autumn.

Habitat.- Open places in the bog and along the margins; more common in sterile, low grounds and pastures; not typically a bog plant but sometimes found there.

Notes.- Attractive in flower and much sought by bees for nectar and pollen; sometimes a nuisance in hayfields and other places where the tough, wiry stems foul up the knives of mowing machines.

Other Names.- Thé du Canada (Que.), Steeple-bush.

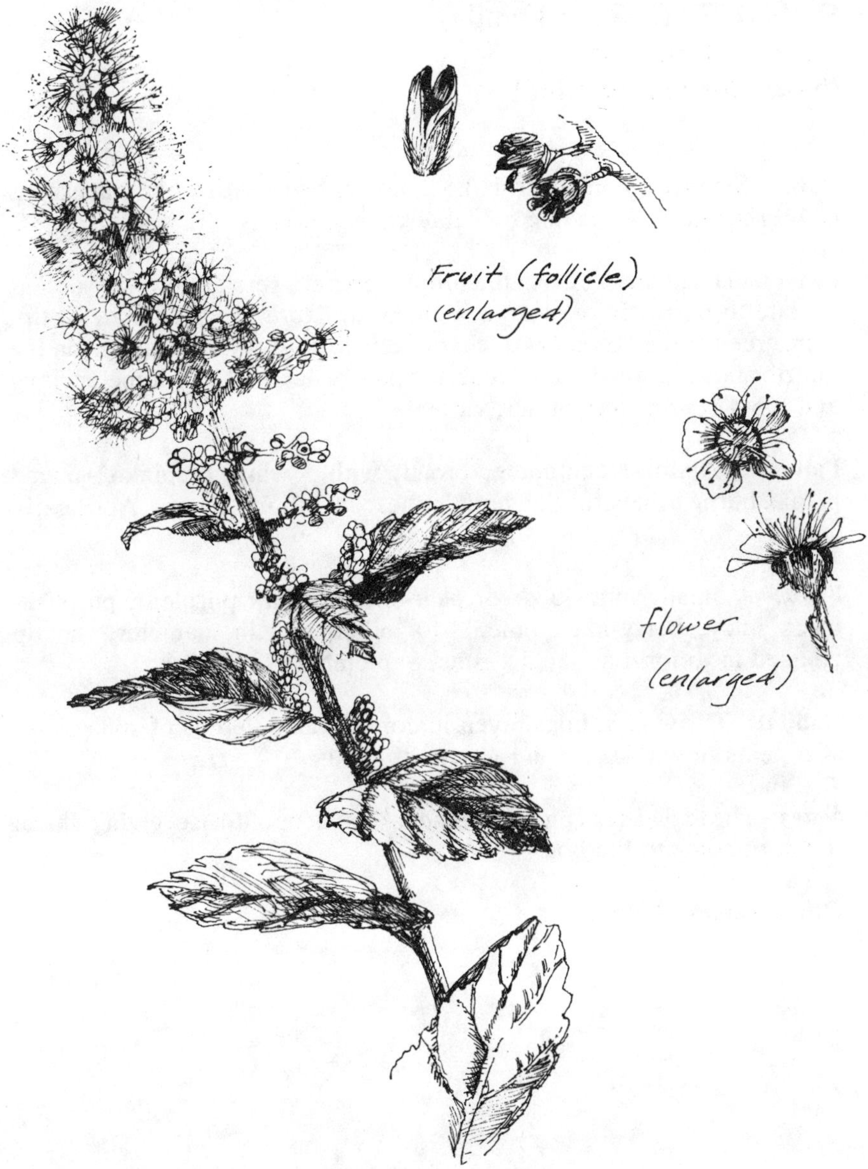

Fruit (follicle)
(enlarged)
flower
(enlarged)

ROSACEAE (Rose Family)
Purple Chokeberry
Pyrus floribunda Lindl.

Habit.- Slender colonial shrub 0.5-3 meters high, with white-pubescent new branchlets, spreading by subterranean offsets.

Leaves.- Deciduous, alternate, simple, crenate-serrate, broadly oblanceolate to narrowly obovate or subelliptic, abruptly acuminate at tip, dark green and glabrous with characteristic stalked, dark glands on the midrib above, grayish tomentose and pale beneath, 2-9 centimeters long and 0.5-4 centimeters broad, variable.

Flowers.- About 1 centimeter broad, with 5 white or pinkish-tinged petals, borne in several-flowered, pubescent inflorescences, April-early July.

Fruit.- A small, subglobose or pear-shaped, dark purple or purplish-black juicy, berry-like pome, 8-10 millimeters in diameter, the tip creased in to form a 5-angled star, September-November.

Habitat.- Thriving in bogs, even in considerable depths of sphagnum; also peats, low thickets and wet to dry soils.

Notes.- The leaves turn pleasing shades of pink or old-rose, giving a local touch of color to the bog.

Other Names.- None.

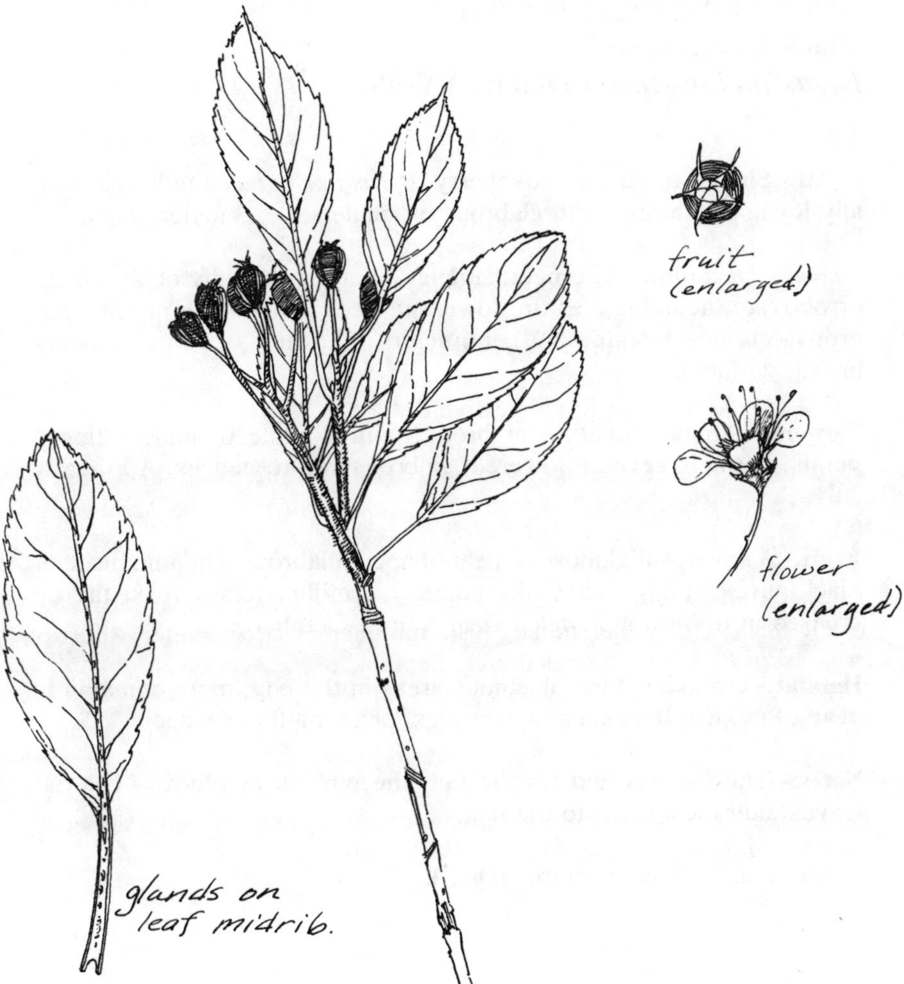
fruit (enlarged)
flower (enlarged)
glands on leaf midrib.

ROSACEAE (Rose Family)
Black Chokeberry
Pyrus melanocarpa (Michx.) Willd.

Habit.- Similar to purple chokeberry (*Pyrus floribunda* Lindl.) but usually lower in stature, with glabrous branchlets, 0.5-3 meters high.

Leaves.- Deciduous, alternate, simple, elliptic to broadly oblanceolate or obovate, acuminate at tip, lower surfaces (except the midrib) glabrous, crenate-serrate, 2-9 centimeters long and 0.5-4 centimeters broad, variable.

Flowers.- About 1 centimeter broad, with 5 white or pinkish-tinged petals, borne in several-flowered, glabrous inflorescences, April-early July.

Fruit.- A small, subglobose or pear-shaped, glabrous, shining, black or black-purple, juicy, berry-like pome, 6-8 millimeters across, the tip creased in to form a 5-angled star, August-October.

Habitat.- Occasional in sphagnous areas in the bog; more common in peats, low, dry thickets and clearings, or on bluffs or ledges.

Notes.- The flowers and (in the fall, the pinkish or old-rose-colored leaves) add local color to the bog.

Other Name.- Gueules noires (Que.).

fruit: Longitudinal Section. (enlarged.)
Glands on leaf midrib.

ROSACEAE (Rose Family)
Shrubby Cinquefoil
Potentilla fruticosa L.

Habit.- Shrub 0.2-1 meter tall with pale, outer shreddy bark.

Leaves.- Deciduous, alternate, odd-pinnately compound with 5 or 7 entire, narrowly oblong to lanceolate or oblanceolate leaflets, with revolute margins, glabrous or densely white-villous, with persistent, sheathing stipules.

Flowers.- 1.5-3 centimeters broad, yellow, with 5 petals, June-October.

Fruit.- Numerous small, dry achenes borne on a conical receptacle.

Habitat.- Occasional in bogs but more common on wet or dry open ground especially of calcareous origin.

Notes.- Many horticultural forms known in the trade.

Other Names.- Golden-Hardhack, Widdy (Nfld.).

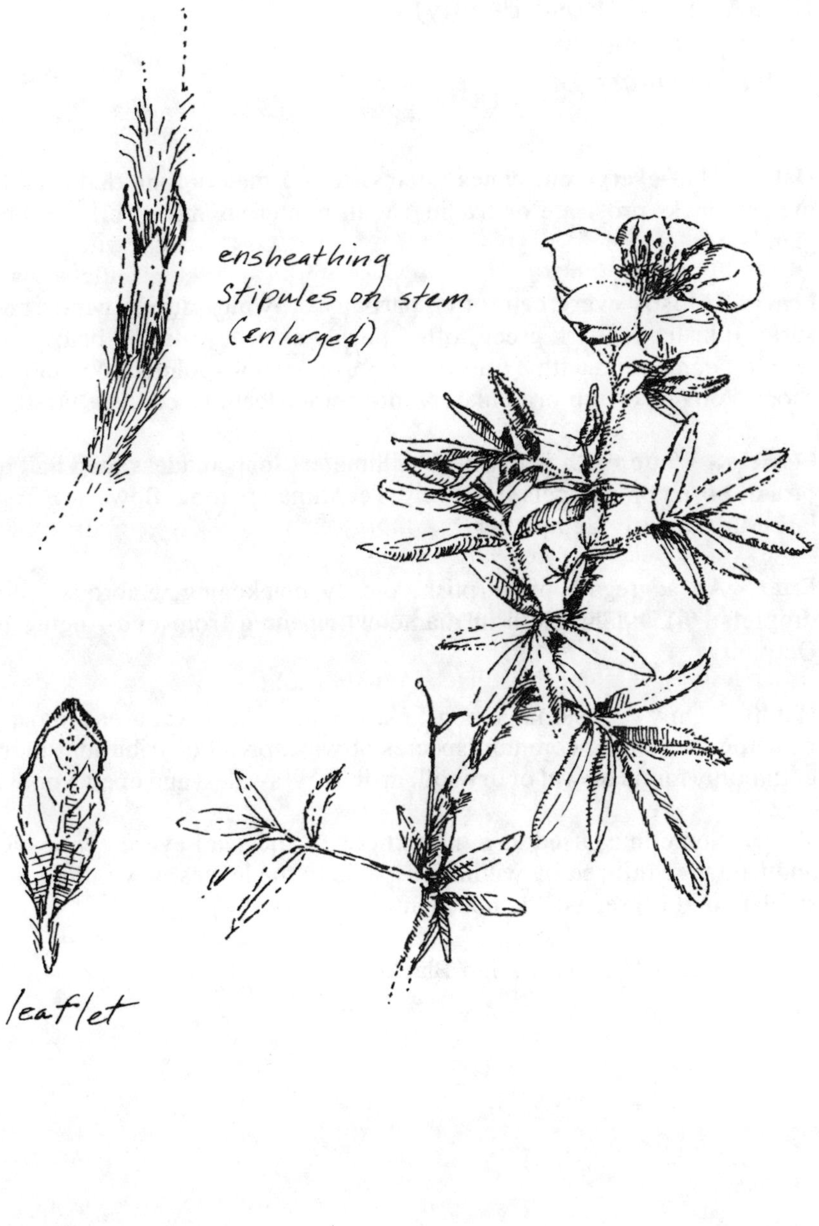
ensheathing
Stipules on stem.
(enlarged)
leaflet

ROSACEAE (Rose Family)
Swamp Dewberry
Rubus hispidus L.

Habit.- Half-evergreen; canes (stems) to 2.5 meters long and 5 millimeters thick, prostrate or trailing, with numerous mixed bristles and glands.

Leaves.- Mostly evergreen (often partially browning during winter and spring), lustrous, dark green, often purple-or bronze-tinted below, alternate, compound with 3 (rarely 5) obtuse or short-pointed, obovate to rhombic-ovate or sub-orbicular, blunt-toothed leaflets; pedicels bristly.

Flowers.- White, with petals 5-12 millimeters long and less than half as broad, borne on pedicels 0.5-rarely 4 centimeters long, flowering from late May to early September.

Fruit.- An aggregate of purplish, tardily blackening, glabrous, sour druplets, 6-15 millimeters in diameter, ripening from mid-August to October.

Habitat.- Only occasional in bogs, usually in the less-wet areas, mostly near the margins; a common species of widespread distribution, more commonly found on wet or dry soil, in ditches, swales, and open woods.

Notes.- Sometimes used as a ground cover; the semi-evergreen leaves and fruits are utilized by wildlife, especially the leaves, in winter when similar food (“greens”) is not plentiful.

Other Names.- Mûrier, Swamp Blackberry

fruit

ROSACEAE (Rose Family)
Northeastern Rose
Rosa nitida Willd.

Habit.- Low, slender, extremely bristly shrub 0.2-1 meter high, with canes (stems) 2-5 millimeters in diameter, from slender, stoloniferous rhizomes.

Leaves.- Deciduous, alternate, odd-pinnately compound with 5-9 finely serrate, narrrowly elliptic or oblong-oval leaflets; stipules adnate to the petiole (a distinctive feature of roses).

Flowers.- Rather conspicuous, pink, solitary or in bristly, few-flowered inflorescences; in evening with fragrance of Lily-of-the-Valley (*Convallaria*), June-August.

Fruit.- Hairy achenes enclosed in a dark red, subglobose, fleshy receptacle or "hip".

Habitat.- Wet, open areas, often near the bog border; also wet thickets, pond margins, etc.

Notes.- Occasional in bogs, lending local color and fragrance.

Other Name.- Swamp Rose.

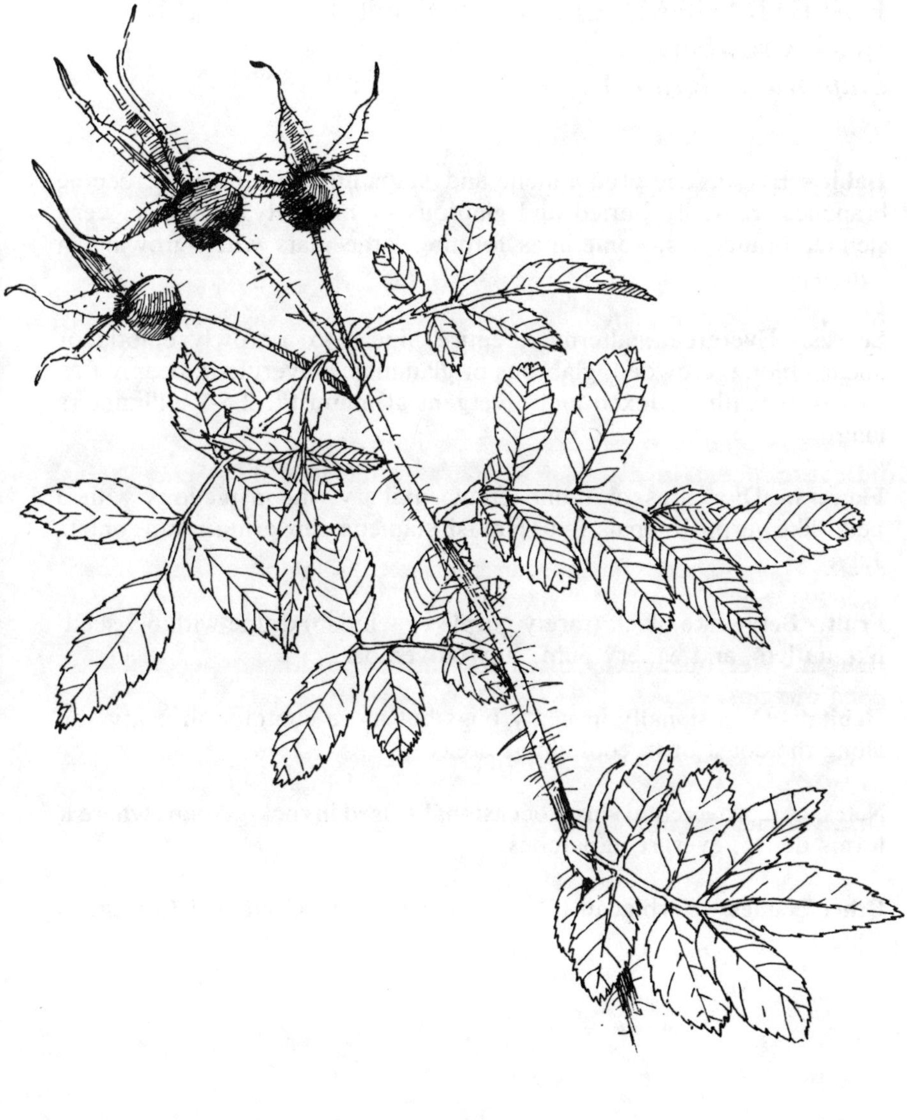

EMPETRACEAE (Crowberry Family)
Black Crowberry
Empetrum nigrum L.

Habit.- Evergreen, procumbent and spreading shrub with creeping branches, partially buried and glabrous or minutely glandular, very slender branchlets, sometimes forming large mats when growing in humus.

Leaves.- Evergreen, alternate, entire, linear to narrowly elliptic or linear-oblong, crowded, glabrous or glandular pulverulent, deeply furrowed beneath, reflexed and divergent at maturity, 2.5-7 millimeters long.

Flowers.- Dioecious or monoecious, axillary, inconspicuous with 3 petal-like purplish sepals and 3 pinkish stamens, flowering in June or late July.

Fruit.- Berry-like black (rarely purple or whitish) drupe with 6-9 seed-like nutlets, and watery pulp, July-November.

Habitat.- Occasionally in peaty bogs but more common in peaty soil along the coast or in cool alpine areas.

Notes.- An ornamental shrub occasionally used in rock-gardens where it forms dense, evergreen patches.

Other Names.- Corbigeau (Que.), Curlewberry, Graines a Corbigeau (Que.).

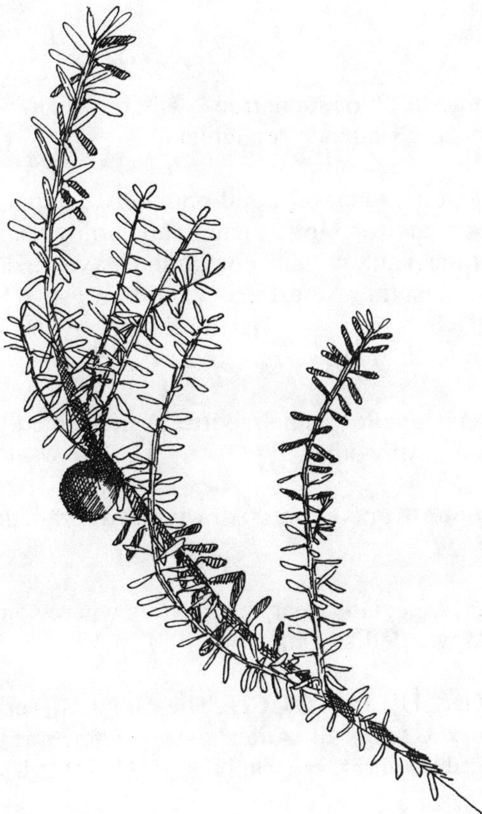

ANACARDIACEAE (Cashew Family)
Poison Sumac
Rhus vernix L.

Habit.- A coarse shrub or small tree 2-7 meters high, with gray, smooth bark and glabrous, glaucous branchlets.

Leaves.- Deciduous, alternate, odd-pinnately compound with 7-13 entire, oblong-obovate to elliptic, acuminate, shiny leaflets. The general appearance of the plant, with its compound leaves and reddish petioles, closely resembles the American Mountain-ash (*Pyrus americana* (Marsh. DC.) but may be distinguished by its *entire* leaflets. (Mountain-ash leaflets are *serrate*).

Flowers.- Greenish-yellow, small, borne in slender axillary panicles 8-20 centimeters long, May-June.

Fruit.- A subglobose or compressed, whitish or drab, dry drupe, August to November.

Habitat.- Bogs, especially near the borders where some mineral soil is available; also wooded swamps.

Notes.- POISONOUS TO TOUCH! The rather attractive whitish fruits may persist on the twigs all winter and are sometimes tempting to the novice for floral displays, but DON'T TOUCH! It is *rash* folly!

Other Names.- Poison-dogwood, Poison-elder.

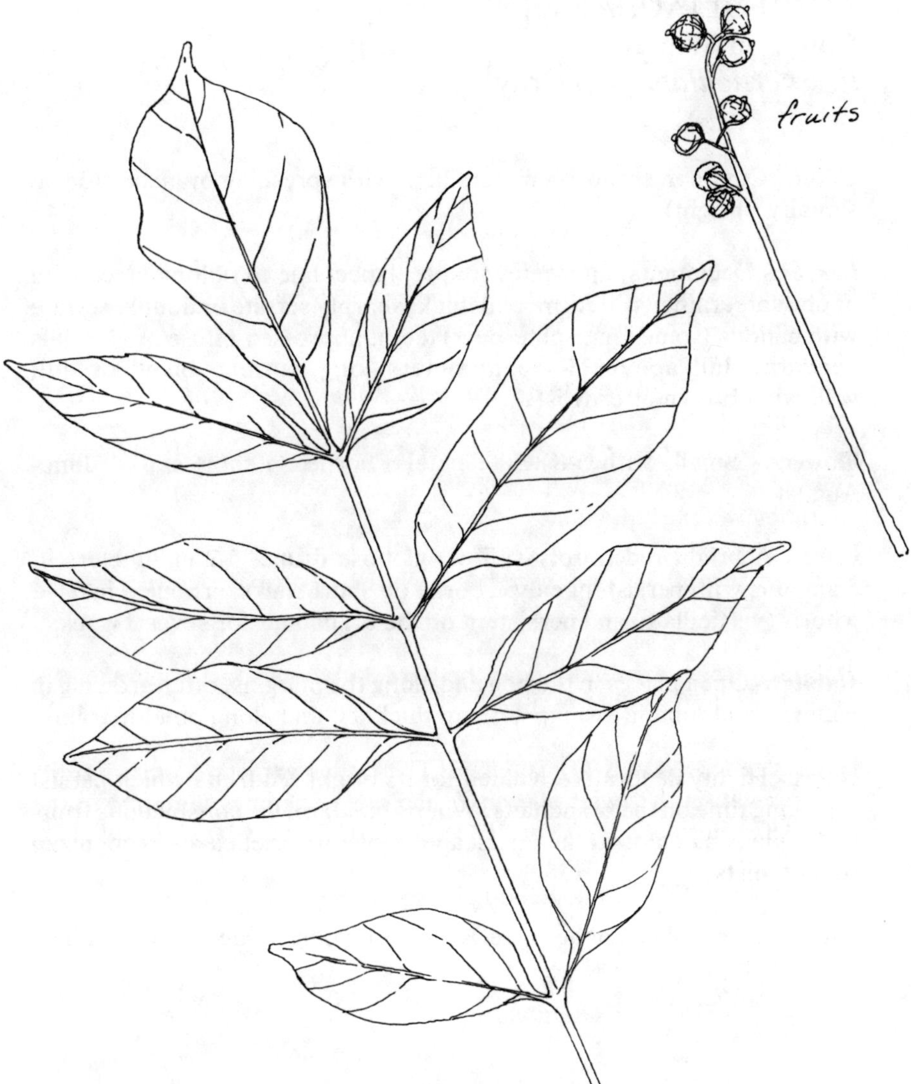
fruits

AQUIFOLIACEAE (Holly Family)
Common Winterberry
Ilex verticillata (L.) Gray

Habit.- Common shrub 1-4 meters high with spreading branches (occasionally upright).

Leaves.- Deciduous, alternate, simple, lanceolate to oblong-lanceolate or obovate, rather stiff, firm, and thick, sharply serrate or doubly serrate with callous (somewhat spinescent) teeth. appressed-pilose or glabrous beneath, dull above, 3-7 centimeters long, turning black or dull yellowish-brown after frost.

Flowers.- Small, with 5-8 white petals, borne on short stalks, June-August.

Fruit.- A bright red (rarely yellow) globose drupe, 5-7 millimeters in diameter, with persistent calyx, borne on short stalks, arranged in false whorls (verticillate) and persistent on the branchlets for several weeks.

Habitat.- Open places in the bog and along the margins, often growing in water; also found in swamps, damp thickets, and along pond-margins.

Notes.- Highly decorative, valued for its bright red fruits which persist for a long time on the branchlets. When abundant, the conspicuous fruits add a pleasing color to the landscape. Several varieties are recognized by botanists.

Other Names.- Apalanche (Que.), Aulne blanche (Que.), Black Alder, Winterberry.

AQUIFOLIACEAE (Holly Family)
Smooth Winterberry
Ilex laevigata (Pursh) Gray

Habit.- Shrub to 4 meters tall, with upright, glabrous branches.

Leaves.- Deciduous, alternate, simple, lance-ovate to lanceolate or oval-elliptic, 3-9 centimeters long, lustrous above, glabrous or pilose on the veins beneath, appressed-serrulate, long acuminate, turning yellow in the fall.

Flower.- Small, with 4 or 5 white petals, borne on short stalks, May-June.

Fruit.- A depressed-globose, scarlet or orange-red (rarely bright yellow) drupe, 7-8 millimeters in diameter, with persistent calyx, solitary, on stalks 2-5 millimeters long, persistent.

Habitat.- Thin, sphagnous areas in the bog but more often in wooded swamps or peaty areas.

Notes.- Sometimes planted for its orange-red fruits. Absent north of southern Maine.

Other Name.- Aulne blanche (Que.).

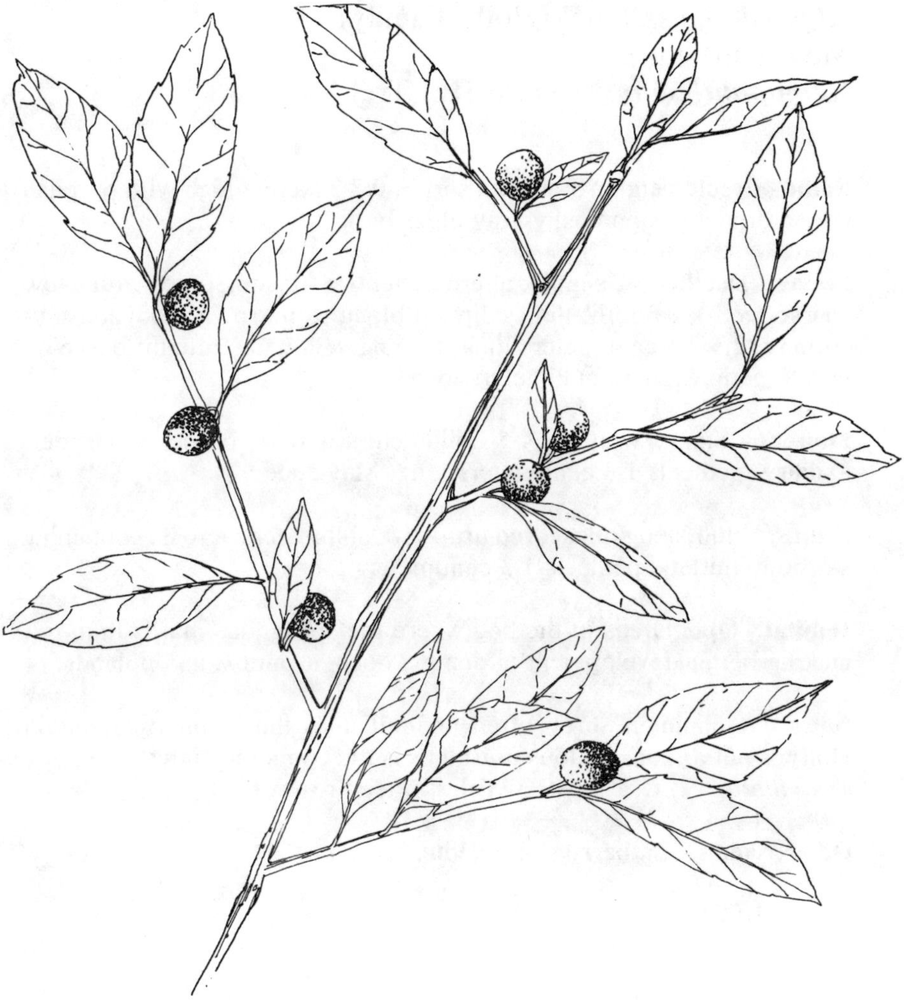

AQUIFOLIACEAE (Holly Family)
Mountain Holly
Nemopanthus mucronata (L.) Trel.

Habit.- Erect, slender-branched shrub, 0.3-3 meters high with purplish young branchlets and ashy-gray older bark.

Leaves.- Deciduous, simple, alternate, entire (or sometimes with a few coarse teeth), smooth, thin, elliptic-oblong to narrowly obovate, mucronate, 2.5-3.5 centimeters long, turning yellow in autumn; borne on purple petioles, 6-12 millimeters long.

Flowers.- Small, whitish, 4-5- millimeters across, borne on slender, axillary, pedicels 1-3 centimeters long, May-June.

Fruit.- A dull, subglobose, red drupe 6-8 millimeters across, containing 4-5 bony nutlets; pedicels 1-3 centimeters long.

Habitat.- Open areas in the bog where the sphagnous mat is thin, but making better development in damp woods, swamps, and uplands.

Notes.- A common shrub often unfamiliar to the layman. Mountain Holly is not as conspicuous in fruit as is the common winterberry, *Ilex verticillata* (L.) Gray, with which it often grows.

Other Names.- Catberry, Faux Houx (Que.).

ACERACEAE (Maple Family)
Red Maple
Acer rubrum L.

Habit.- Common forest tree with irregular or rounded crown, to 40 meters tall (much stunted in bogs); branchlets and twigs red.

Leaves.- Deciduous, opposite, simple, 3-5 palmately lobed, with sharp sinuses coarsely serrate nearly to the base, 6-10 centimeters long, turning brilliant shades of red to yellow in the fall (or even mid-summer), in sharp contrast with the green background of pine and spruce, enhancing the bog with unforgettable beauty.

Flowers.- Dark red to scarlet, appearing in early spring long before the leaves, sometimes injured by late spring frosts, March-May.

Fruit.- Small, paired samaras (''keys'') with pinkish or red wings, maturing early May to July.

Habitat.- Borders or open places in the bog; a common tree of varied habitats, including swamps and uplands.

Notes.- Often planted as a street tree. The scarlet leaves add much color to the landscape in the fall.

Other Names.- Plaine (Que.), Plaine Rouge (Que.), Scarlet Maple, Soft Maple, Swamp Maple, Water Maple, White Maple.

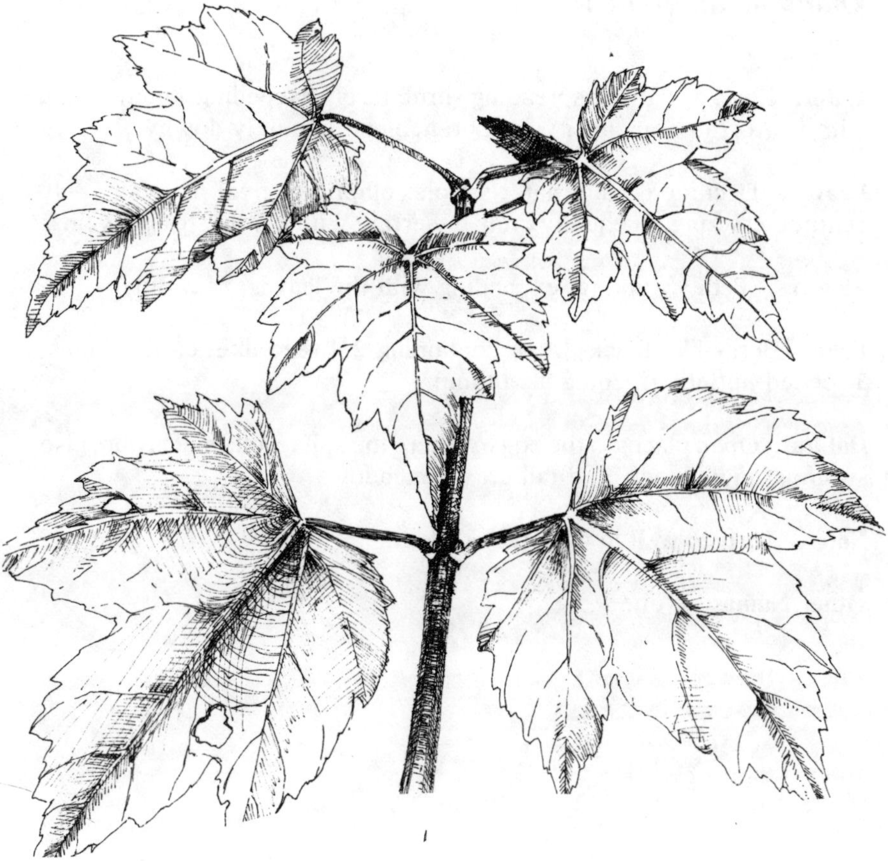

RHAMNACEAE (Buckthorn Family)
Alder-leaved Buckthorn
Rhamnus alnifolia L'Her.

Habit.- Low, upright or spreading shrub rarely exceeding 1 meter high, with few or no branches; young branchlets, minutely downy.

Leaves.- Deciduous, alternate, simple, elliptic to oval or ovate, 4-10 centimeters long, unequally crenate-serrate with nearly straight veins.

Flowers.- Petals wanting, expanding with the leaves, May-July.

Fruit.- Berry-like black drupe containing 2-4 seed-like, cartilaginous, 3-seeded nutlets, ripening in autumn.

Habitat.- Open places in the bog or where the sphagnous mat is thin; also swamps, low woods, and tall-grassy meadows.

Notes.- Thrives well in calcareous areas.

Other Names.- None.

LYTHRACEAE (Loosestrife Family)
Water-willow
Decodon verticillatus (L.) Ell.

Habit.- Aquatic smooth or downy subshrub with angled, recurved stems, 4-6 decimeters long, the arching stems rooting at the tips; bark of submersed stems conspicuously spongy-thickened.

Leaves.- Deciduous, simple, opposite or whorled, lanceolate, nearly sessile, entire with undulate margins, 3-20 centimeters long.

Flowers.- Borne on short pedicels in clusters of 2-8 in the axils of the upper leaves; petals magenta or rose-purple, 8-12 millimeters long, July-August.

Fruit.- A 3-5 valved, subglobose capsule 5 millimeters in diameter, bearing angular seeds.

Habitat.- Open places in the bog, growing in water; also swamps and shallow pools.

Notes.- Occasional in bogs: unique because of its excessive development of soft spongy cork on the submersed parts of stems.

Other Name.- Water Oleander.

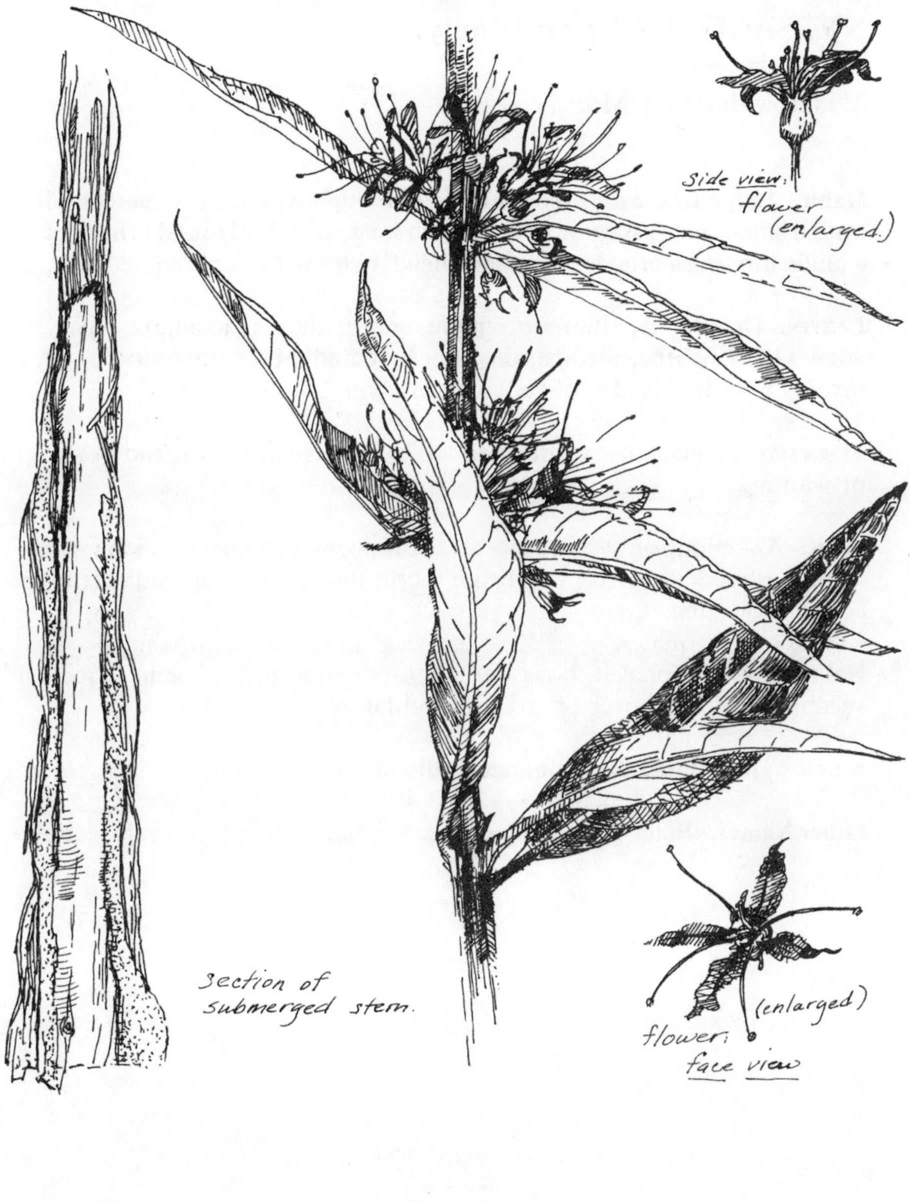
Side view: flower (enlarged.)
Section of submerged stem.
(enlarged)
flower: face view

NYSSACEAE (Tupelo Family)
Black Tupelo
Nyssa sylvatica Marsh.

Habit.- Large tree with fissured bark and tough wood, to 30 meters tall (much smaller in sphagnous bogs), with long, stiff, horizontal branches, usually forming a broad, flat-topped head; twigs with diaphragmed pith.

Leaves.- Deciduous, alternate, simple, entire, obovate to elliptic, shiny, somewhat leathery, 3-10 centimeters long and 2-6 centimeters broad, turning pleasing shades of red and yellow in autumn.

Flowers.- Inconspicuous, greenish-white; petals very small and fleshy, or wanting.

Fruit.- An oblong or ovoid, blue-black drupe, 8-12 millimeters long with a 10-12-ribbed stone (pit), and thin, acrid flesh, borne on pedicels 3-6 centimeters long.

Habitat.- Occasional in bogs but more common in low, acid woods, swamps, and on shores of streams and lakes.

Notes.- The bright scarlet foliage is distinctive in autumn.

Other Names.- Black gum, Pepperidge, Sour gum, Tupelo, Tupelo gum.

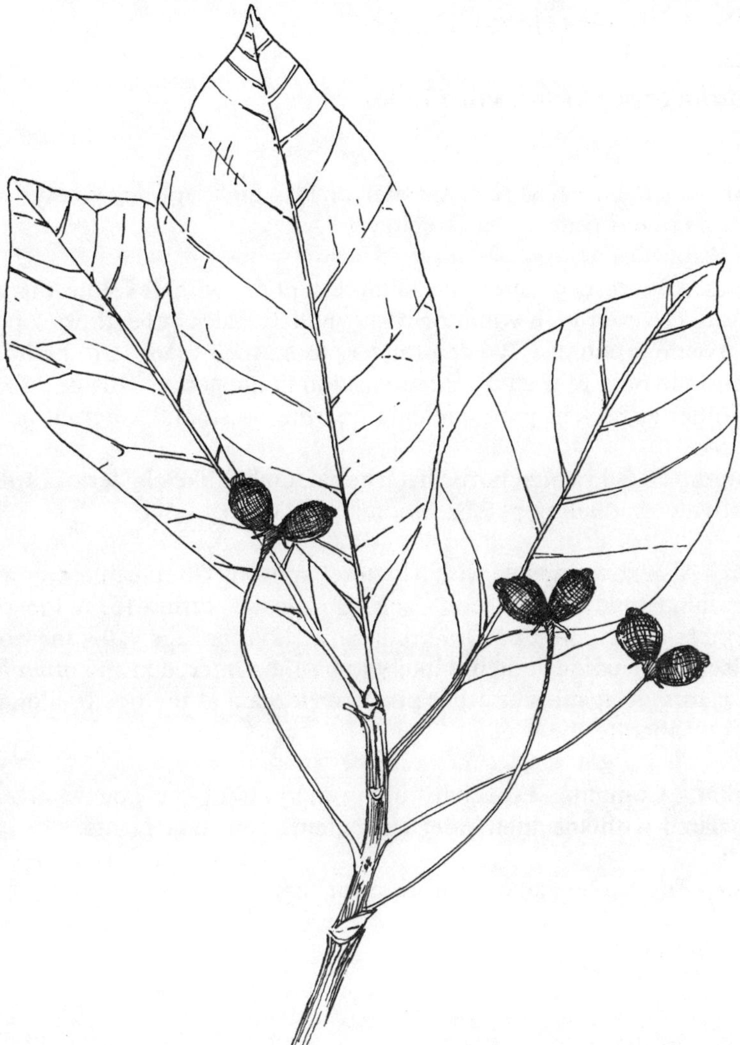

ERICACEAE (Heath Family)
Labrador Tea
Ledum groenlandicum Oeder

Habit.- Low boreal shrub, 1 meter or less high, rooting freely in the sphagnum and peat.

Leaves.- Evergreen, alternate, simple, entire, with revolute margins, densely covered with white or rusty, woolly hairs beneath, oblong, or linear-oblong, obtuse, 2-5 centimeters long, their blades often directed toward the base of the twig due to a bend in the petiole. Bruised foliage and other succulent parts yielding a strong, resinous, fragrant aroma.

Flowers.- Small, white, borne in terminal, umbel-like clusters; corolla of 5 obovate, spreading petals, blooming in May or June.

Fruit.- A slender capsule with 5 locules, splitting from the base upward, containing many tiny seeds. The capsules, splitting from the base, resemble a partially open umbrella or a shooting star. This method of dehiscence is unique and not likely to be encountered in any other local bog plants. (Capsules of other plants split open at the tips or along the sides to liberate the seeds.)

Habitat.- Common, especially along margins of the bog where it is associated with speckled alder and other tall woody plants.

Notes.- The leaves have been used for tea.

Other Names.- None.

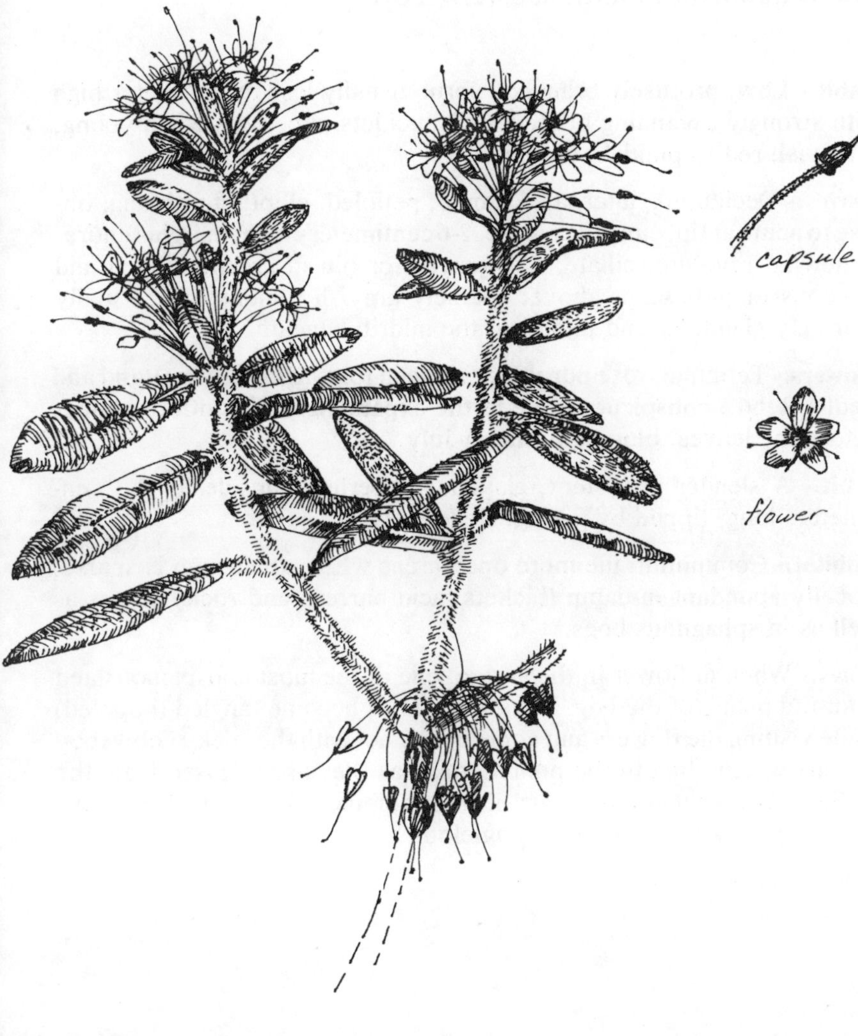
capsule
flower

ERICACEAE (Heath Family)
Rhodora
Rhododendron canadense (L.) Torr.

Habit.- Low, profusely branched shrub usually less than 1 meter high with strongly ascending branches. Branchlets puberlous when young, yellowish red to pinkish, bloomy.

Leaves.- Deciduous, alternate, simple, petioled, elliptic to oblong, obtuse to acute at tip, cuneate at base, 2-6 centimeters long; margins entire, somewhat revolute, ciliate; dull grayish or bluish green, shining and glaucous or pubescent above; sparsely grayish tomentulose, usually sparingly glandular and pilose on the midrib beneath.

Flowers.- Terminal, rose purple, or white in forma *albiflorum* (Rand and Redf.) Rehd., conspicuous due to the expanding of the inflorescences before the leaves, blooming March-July.

Fruit.- A slender persistent, glaucous-puberlent capsule 0.7-1.5 centimeters long, tipped by a long, slender style.

Habitat.- Common in the more open areas where sphagnum is sparse. Locally abundant in damp thickets, acid barrens and rocky slopes as well as in sphagnous bogs.

Notes.- When in flower in the spring, one of the most conspicuous and beautiful plants of the bog. Bees sometimes become tangled (hobbled) while visiting the flowers and come in contact with the sticky cobwebby threads which cling to the pollen grains as they are released from the anthers. Bee keepers lose part of their colonies each year due to this phenomenon known as ''bee tanglefoot''.

Other Names.- None.

ERICACEAE (Heath Family)
Lambkill
Kalmia angustifolia L.

Habit.- Slender evergreen shrub up to 1.7 meters high. Branchlets terete, strongly ascending, glabrous or nearly so, with naked buds.

Leaves.- Evergreen, simple, opposite or ternate, entire, flat, rather thin, oblong to elliptic-lanceolate, glabrous or puberulent and glabrate, (ferruginous when young), on short petioles, obtuse or acutish, bright green above, paler beneath; blades often bending down due to the curved petioles.

Flowers.- Deep rose-pink or purplish crimson (rarely white), borne laterally on slender pedicels along the stem; pedicels glandular-puberulent, recurving in fruit, blooming in late May-August.

Fruit.- A rather small, thin-valved, fragile, depressed-globose capsule.

Habitat.- A small shrub commonly found on wet sterile soil, old pastures and barrens (often a weed in blueberry fields), but commonly invading and surviving in sphagnous bogs.

Notes.- Leaves poisonous to livestock.

Other Names.- Dwarf Laurel, Pig Laurel, Sheep Laurel, Wicky.

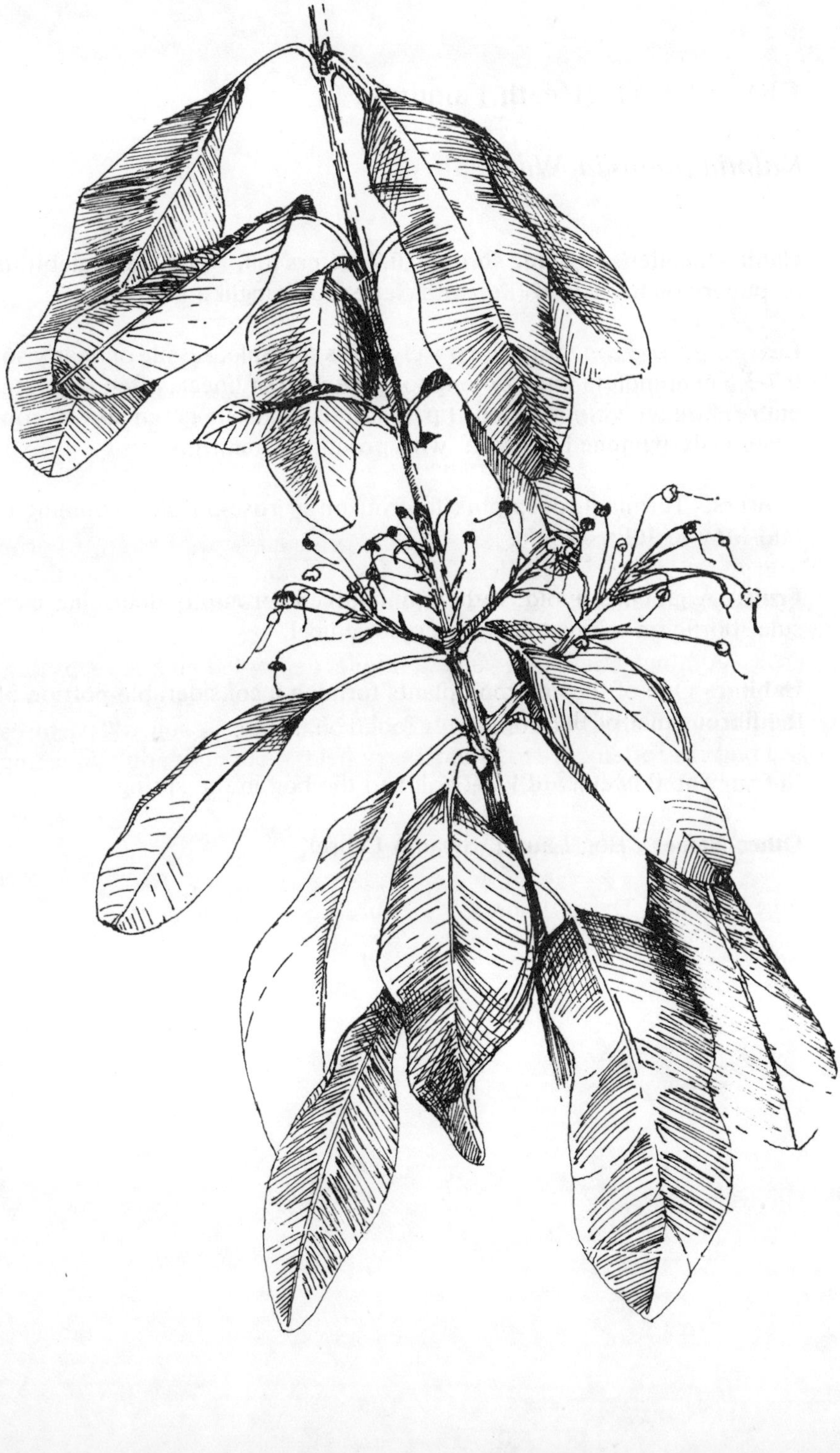

ERICACEAE (Heath Family)
Pale-Laurel
Kalmia polifolia Wangenh.

Habit.- Slender straggling shrub to 0.7 meters tall. Branchlets, glabrous or puberulous, 2-edged (i.e. with flattened, winglike margins).

Leaves.- Evergreen, opposite (rarely in 3's), simple, sessile or nearly so, 0.7-3.5 centimeters long, oblong to lanceolate or linear, often revolute, entire, firm with blunt callous tip, glabrous, lustrous-green above, conspicuously whitened beneath, with prominent midrib.

Flowers.- Terminal, deep pink to crimson or rose-purple, blooming in mid-May to July.

Fruit.- A globose-ovoid, rather thin-valved, promptly-dehiscing capsule, borne on a long, slender, erect pedicel.

Habitat.- One of the common plants forming a considerable portion of the fibrous mat of the bog.

Notes.- The flowers add local color to the bog in the spring.

Other Names.- Bog Laurel, Swamp Laurel.

fruit
(enlarged)
(capsule)

ERICACEAE (Heath Family)
Bog-Rosemary
Andromeda glaucophylla Link

Habit.- Low, pale, evergreen shrub with elongate creeping base and ascending, nearly terete, little-branched, glaucous stems 1-7 decimeters high.

Leaves.- Evergreen, alternate, simple, firm, revolute, linear to narrowly oblong, white-tomentulose beneath.

Flowers.- White, terminal, nodding, May-July.

Fruit.- A depressed- globose, glaucous capsule with rather thick valves, borne on short, thick, recurved pedicels.

Habitat.- Abundant, its creeping root-stalks composing a considerable portion of the fibrous mat of peat bogs, able to thrive where many other woody plants could not survive in the thick accumulation of living sphagnum; also common in peat, on margins of pools and other continuously wet places where it may not be associated with sphagnum.

Other Name.- Andromeda.

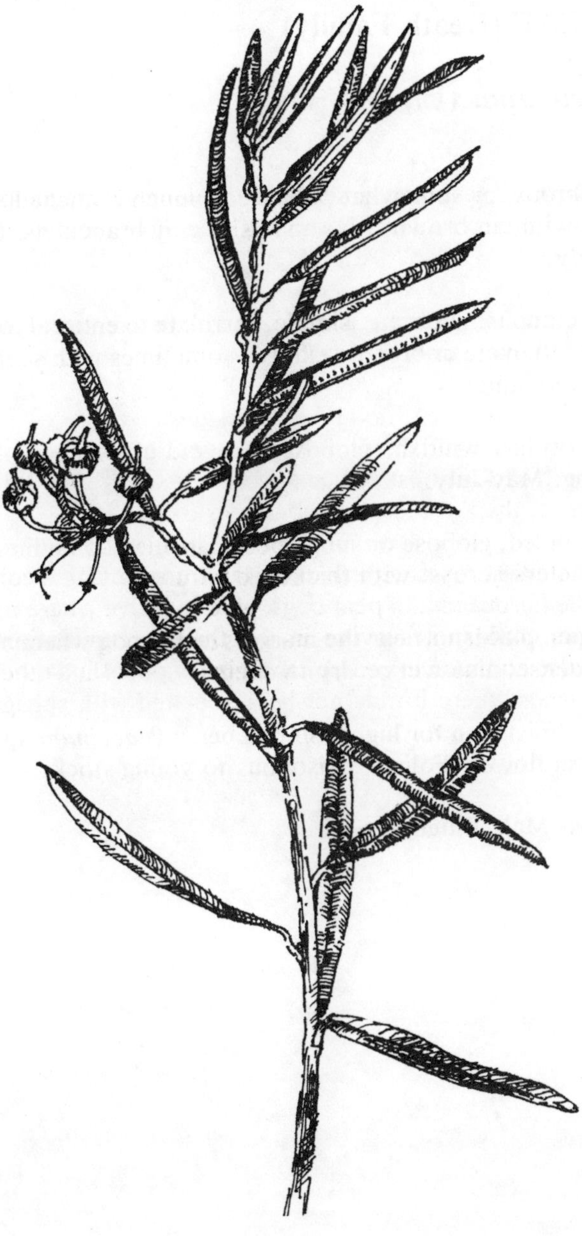

ERICACEAE (Heath Family)
Maleberry
Lyonia ligustrina (L.) DC.

Habit.- Glabrous, or somewhat pubescent, much-branched shrub to 4 meters tall, with tan-brown twigs and ashy gray branchlets; outer bark thin, shreddy.

Leaves.- Deciduous, alternate, simple, serrulate to entire, lanceolate or oblanceolate to ovate or broadly elliptic, sometimes with scurfy scales, 3-7 centimeters long.

Flowers.- Corollas whitish, globose-or ovoid-urn-shaped, 2.5-5 millimeters long, May-July.

Fruit.- A 5-angled, globose or subglobose capsule, 2-5 millimeters long and 3 millimeters across, with thickened sutures.

Habitat.- Open places or near the margins of the bog where the sphagnum is not dense; also wet or dry thickets.

Notes.- Often mistaken for highbush blueberry (*Vaccinum* spp.), especially when in flower. Foliage poisonous to young stock.

Other Name.- Male Blueberry.

fruits. (enlarged)

ERICACEAE (Heath Family)
Leather-leaf
Chamaedaphne calyculata (L.) Moench

Habit.- Upright, low, much-branched, nearly evergreen, circumpolar shrub to 1.5 meters tall, with scaly buds and slender twigs; branchlets slender, spreading or horizontal, tan-brown, appearing white-striped by numerous, linear or narrowly elliptic white lines due to exfoliating of the thin, outer, waxy or scurfy layer.

Leaves.- Nearly evergreen, alternate, simple, entire (often with wavy margins), revolute, oblong to oblong-lanceolate, 2.5-5 centimeters long and one-third as wide (becoming progressively smaller toward the tip of the twig), dull green and slightly scaly above, densely scurfy below.

Flowers.- Axillary; corolla white, ureolate, 6-7 millimeters long, March-July.

Fruit.- Capsules depressed, many-seeded, 4 millimeters across, tipped by a persistent style (resembling a pin stuck in a pin cushion).

Habitat.- Abundant in bogs, especially near the border where the sphagnum is less dense; also common in peaty swales, pond margins, etc., where water is plentiful.

Notes.- Eagerly visited by bees early in the spring because the flowers appear before those of most other plants, and bees take advantage of the early treat.

According to *Gray's Manual of Botany,* ed. 8 (1950), by M.L. Fernald, *Chamaedaphne calyculata* (L.) Moench is a Eurasian species and not part of our flora. It is, however, represented by two varieties as follows:

var. *angustifolia* (Ait.) Rehd., and

var. *latifolia* (Ait.) Fern. Our illustration and accompanying description will be general enough to include these two varieties.

Other Name.- Cassandra.

section of twig
with
fruits(enlarged)

ERICACEAE (Heath Family)
Creeping Snowberry
Gaultheria hispidula (L.) Bigel.

Habit.- Trailing and creeping, delicate, matted evergreen plant with slightly woody stems and coarsely appressed setose, yellowish-brown or tan colored branchlets.

Leaves.- Evergreen, alternate, simple, entire, orbicular-ovate to ovate, narrowed at the ends, revolute, lustrous and glabrous above, paler beneath and rusty-strigose on midrib, 4-10 millimeters long.

Flowers.- 4 millimeters long, white, mostly solitary in the axils on short, nodding peduncles, April or May (or until August, in the mountains).

Fruit.- A shiny, white, subglobose, many-seeded capsule (enclosed when ripe by a fleshy calyx so as to appear as a globular, slightly bristly, juicy, slightly acid, aromatic (wintergreen-tasting) berry), 6 millimeters in diameter.

Habitat.- Forming dense, bright green carpets in the bog, studded with white "berries" in summer; also in mossy coniferous woods and uplands.

Notes.- Formerly known as *Chiogenes hispidula* Torr. & Gray.

Other Names.- Capillaire, Maidenhair-berry, Moxieplum, Oeufs de Perdrix (Que.), Petit Thé (Que.).

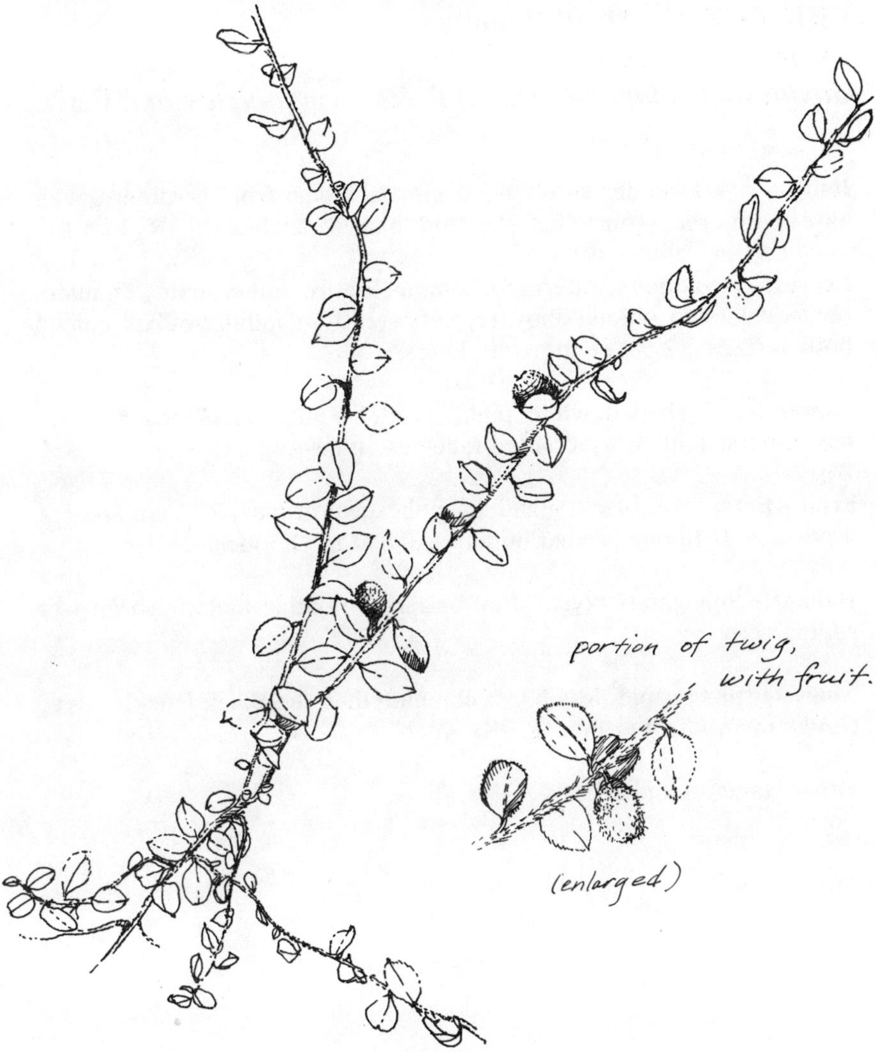
portion of twig, with fruit.
(enlarged)

ERICACEAE (Heath Family)
Dwarf Huckleberry
Gaylussacia dumosa (Andr.) T. & G. var. *bigeloviana* Fern.

Habit.- Low, slender shrub 1-5 decimeters high from a subterranean base, with erect, somewhat glandular-hairy branches.

Leaves.- Deciduous, alternate, simple, entire, mucronate, cuneate-oblanceolate to oblong-obovate, persistently glandular-pubescent on both surfaces, 2.5-4 centimeters long.

Flowers.- Bell-shaped, white, pink, or red, 8-9 millimeters long, borne in loose, persistent, leafy-bracted racemes, June-August.

Fruit.- Berry-like, black, glandular-pubescent drupes, 8-12 millimeters across, with 10 one-seeded nutlets, August to October.

Habitat.- Sphagnous bogs, often on large, floating mats; also on wet peats.

Notes.- Fruits insipid. Much less common than the Black Huckleberry, *Gaylussacia baccata* (Wang.) K. Koch.

Other Name.- Gopherberry.

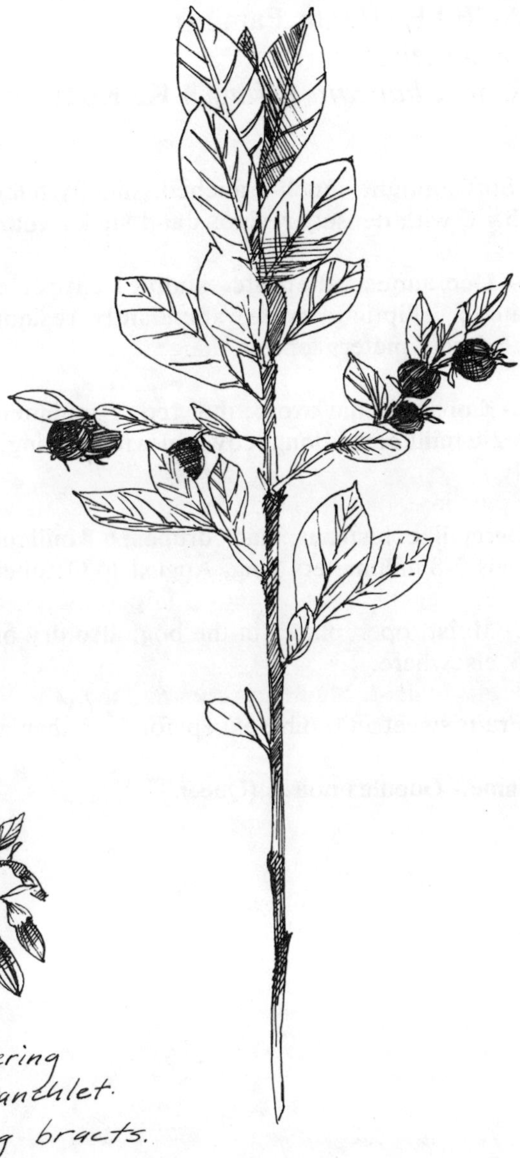

flowering branchlet showing bracts.

ERICACEAE (Heath Family)
Black Huckleberry
Gaylussacia baccata (Wang.) K. Koch

Habit.- Stiff, upright, much-branched, slightly pubescent shrub, 0.3-1 meters high, with densely resinous and sticky young growth.

Leaves.- Deciduous. alternate, simple, entire, elliptic to oblong-lanceolate or elliptic-obovate, abundantly resinous-dotted on both sides, 2.5-5 centimeters long.

Flowers.- Corolla conic-ovoid, dull red, 5 millimeters long, borne on pedicels 2-8 millimeters long, covered with shining, resinous globules, May-July.

Fruit.- Berry-like, lustrous, black drupes, 6-8 millimeters across, borne on pedicels 2-8 millimeters long, August to October.

Habitat.- Moist, open places in the bog; also dry or moist woods and clearings elsewhere.

Notes.- Fruits sweet and edible except for the rather large, nutlike seeds.

Other Name.- Gueules noires (Que.).

longi. section: fruit (enlarged)

ERICACEAE (Heath Family)
Velvet-leaf Blueberry
Vaccinium myrtilloides Michx.

Habit.- Low, much-branched, clone-forming plant, 2-9 decimeters high with minute, warty ("goose-pimply"), fine, wiry, densely-velvety branchlets.

Leaves.- Deciduous, alternate, simple, narrowly-elliptic to oblong-lanceolate, entire, downy beneath and often also above (appearing bluish-green or sage-like), 2-5 centimeters long and 0.5-2.5 centimeters broad, half grown at flowering time.

Flowers.- In dense, terminal clusters; corolla globose-urceolate or short-campanulate, 4-6 millimeters long, greenish or tinged with purple, early spring.

Fruit.- An edible, blue (whitish, in forma *chiocarpum* (Deane) Fern.), rather sour berry 7-10 millimeters in diameter, with a heavy bloom, July-September.

Habitat.- Occasional in the bog but more abundant in moist, open woods, swamps and clearings, including dry, sandy barrens where it is harvested along with other commercial lowbush species.

Notes.- Fruits less sweet and often smaller than those of other species; plants appearing more "leafy" than other species, thus partially obscuring the berries.

Other Names.- Bleuet (Que.), Bluet (Que.), Canada Blueberry, Sour-top Blueberry.

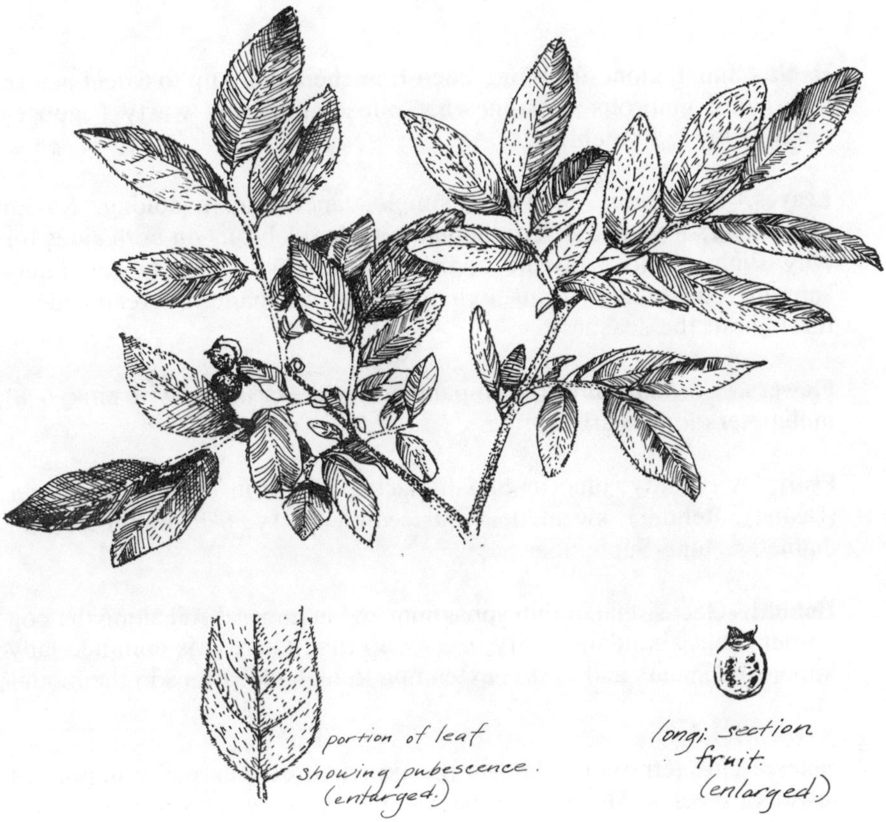
portion of leaf
showing pubescence.
(enlarged.)
longi. section
fruit.
(enlarged.)

ERICACEAE (Heath Family)
Low Sweet Blueberry
Vaccinium angustifolium Ait. var. *laevifolium* House

Habit.- Small, clone-forming, much-branched shrub up to 6 decimeters high, with glabrous or somewhat pilose, minutely warty (''goose-pimply'') new branchlets.

Leaves.- Deciduous, alternate, simple, lanceolate to oblong, 1.5-3.5 centimeters long and 8-15 millimeters wide, glabrous on both sides (or only slightly pilose on the midrib beneath, with closely and finely spinulose-serrulate margins, exhibiting various shades of green and red throughout the season.

Flowers.- Corolla cylindric-campanulate, white or tinged with pink, 6-10 millimeters long, early spring.

Fruit.- A bloomy, blue or bluish-black (whitish in forma *leucocarpa* (Deane), Rehder), sweet delicious, edible berry, 6-15 millimeters in diameter, June-September.

Habitat.- Occasional in thin sphagnum and in mineral soil along the bog borders, but abundant in dry, open barrens (where it is commercially important), peats and rocks, extending to high elevations in the mountains.

Notes.- This tetraploid plant is the principal commercially important lowbush species. The diploid species (*Vaccinium angustifolium* Ait.) is a small, narrow-leaved plant rarely encountered in bogs, and is not here considered. The flowers of all species of blueberries are eagerly visited by bees (upon which the blueberry growers depend for pollination and an ultimate successful harvest of fruits), for pollen and nectar.

Other Names.- Bleuet (Que.), Bluet (Que.), Lowbush Blueberry, Late Sweet Blueberry, Sweet Hurts.

ERICACEAE (Heath Family)
Black-fruited Blueberry
Vaccinium angustifolium Ait. var. *nigrum* (Wood) Dole

Habit.- Similar in stature and size and shape of leaves to var. *laevifolium* House, but with firm, blue-green leaves, glaucous beneath. Clones of this variety, growing along with var. *laevifolium* House occurring in the blueberry barrens, stand out distinctly because of the sharp contrast in leaf color.

Flowers.- Similar to those of var. *laevifolium* House.

Fruit.- Similar to that of var. *laevifolium* House except that the berries are black and without a bloom.

Notes.- The edible fruits are harvested commercially along with other lowbush species with which they grow.

Other Names.- Bleuet (Que.), Bluet (Que.), Dwarf Blueberry.

ERICACEAE (Heath Family)
Highbush Blueberry
Vaccinium corymbosum L.

Habit.- Shrub to 4 meters tall, with stiff, spreading, somewhat gnarled or distorted branches, forming compact or open clumps; young branchlets yellow-green, minutely warty ("goose-pimply").

Leaves.- Deciduous, alternate, simple, ovate to elliptic-lanceolate, entire (or finely serrulate in vars. *albiflorum* (Hook.) Fern., and *glabrum* Gray), glabrous, half-grown at flowering time, 4-8 centimeters long and 2-4 centimeters broad, the lower surface pubescent on the veins, green on both sides (or glaucous beneath in var. *glabrum* Gray).

Flowers.- In dense clusters, corolla white (sometimes pink-tinged), ovoid to cylindric-urceolate, 6-12 millimeters long and 4-6 millimeters wide, early spring.

Fruit.- A blue-black, bloomy, many-seeded, edible (sweet and juicy) berry 6-12 millimeters in diameter, late June to early September.

Habitat.- Mostly in open places or along the wet borders of the bog; also in swamps, low woods, and occasionally in rocky or dry uplands, sparse or absent northward.

Notes.- An interesting shrub because of its delicious fruits, and foliage (which turns bright scarlet and orange in the fall). Also, as in other species of blueberries, much visited by bees for nectar and pollen. Unsightly "witches' brooms" are often present on the branches.

Other Names.- Bleuet (Que.), Bluet (Que.), Swamp Blueberry.

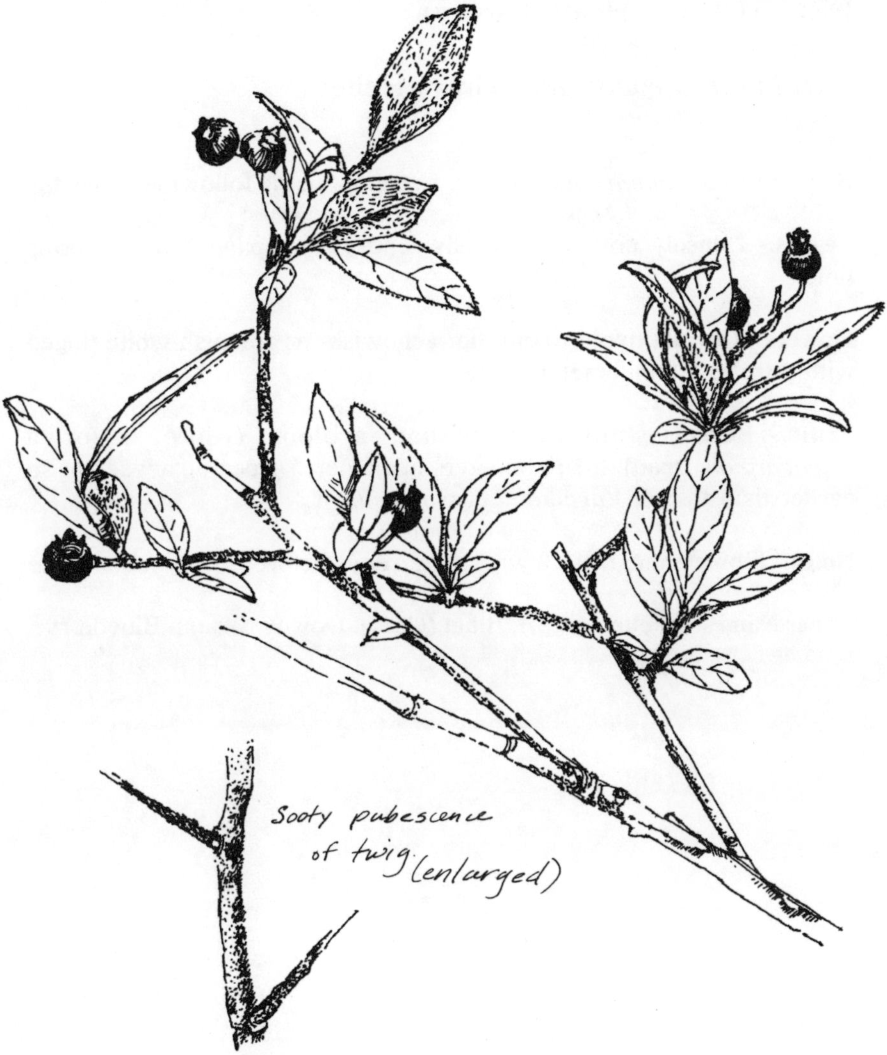
Sooty pubescence
of twig (enlarged)

ERICACEAE (Heath Family)
Black Highbush Blueberry
Vaccinium atrococcum (Gray) Heller

Similar to *Vaccinium corymbosum* L. except in the following respects:

Leaves.- Densely downy or woolly, entire, unexpanded at flowering time.

Flowers.- Corolla ovoid to elliptic, yellowish- or greenish-white tinged with purple, 5-8 millimeters long.

Fruit.- Berries shiny black without a bloom (white, in forma *leucocarpum* Dean), 5-8 millimeters in diameter, ripening a week or so earlier than that of *Vaccinium corymbosum* L.

Notes.- Flowers and fruits a week to ten days before *V. corymbosum* L.

Other Names.- Bleuet (Que.), Bluet (Que.), Downy Swamp Blueberry.

fruit: longi. section
(enlarged)

ERICACEAE (Heath Family)
Small Cranberry
Vaccinium oxycoccus L.

Habit.- Stems extremely slender and thread-like with ascending tips, the branchlets creeping and rooting in the sphagnum.

Leaves.- Evergreen, alternate, simple, ovate-oblong to ovate or triangular, 3-8 millimeters long, 1-3 millimeters broad, strongly revolute, conspicuously whitened or glaucous beneath.

Flowers.- Corolla-segments roseate, 5-6 millimeters long, pedicels 1-4, arising from a terminal short rachis at most 3 or 4 millimeters long.

Fruit.- Red, (often pale and speckled when young) acid, 5-8 millimeters in diameter, globose to pyriform, borne on a long, filamentous pedicel, holding over winter, August-October.

Habitat.- Open places in the bog, where it branches and roots freely; also peaty and upland soils.

Notes.- Berries sour, used for jellies and preserves.

Other Name.- Atocas (Que.).

ERICACEAE (Heath Family)
Large Cranberry
Vaccinium macrocarpon Ait.

Habit.- Stems slender and creeping, elongate and intricately forking, the flowering branches ascending, rooting in the sphagnous mat.

Leaves.- Evergreen, alternate, simple, oblong-elliptic, blunt at tip, 6-18 millimeters long, 2-8 millimeters broad, pale or slightly whitened beneath, flat or slightly revolute.

Flowers.- Corolla-segments 6-10 millimeters long, roseate; pedicels 1-10, arising laterally from an elongate rachis 1-3 centimeters long.

Fruit.- Red, (often pale and speckled when immature), acid, globose, ellipsoid, obovoid or pyriform berry 1-2 centimeters in diameter, September-November, borne on a long, filamentous pedicel, holding through winter.

Habitat.- Open places in the bog where it roots in the sphagnum; also found in swamps on wet shores and in meadows.

Notes.- Cultivated for its berries which are used extensively in making jellies and preserves.

Other Names.- Atocas (Que.), American Cranberry.

OLEACEAE (Olive Family)
Black Ash
Fraxinus nigra Marsh.

Habit.- Sparsely branched, small to medium-sized timber-tree, 15-25 meters tall (smaller in bogs), with rather flaky or corky bark.

Leaves.- Deciduous, opposite, odd-pinnately compound with 7-13 leaflets (the lateral ones sessile); a pad of tawny hairs extending across the rachis at the juncture of the leaflets; bruised foliage with the odor of *Sambucus* (Elder).

Flowers.- Polygamous or dioecious, naked, appearing in long panicles before the leaves emerge, April-May.

Fruit.- Single samara, with nearly flat seed body completely surrounded by the wing.

Habitat.- Wet borders of bogs; also swamps and shores as well.

Notes.- Generally less important as a timber-tree than other northern ashes because of the poorly shaped bole and inferior quality of the wood. However, from early times the wood has been used by the Indians in making pack baskets and other articles in the northeastern United States and Canada.

Other Names.- Basket Ash, Brown Ash, Frêne noir (Que.), Hoop Ash, Swamp Ash, Water Ash.

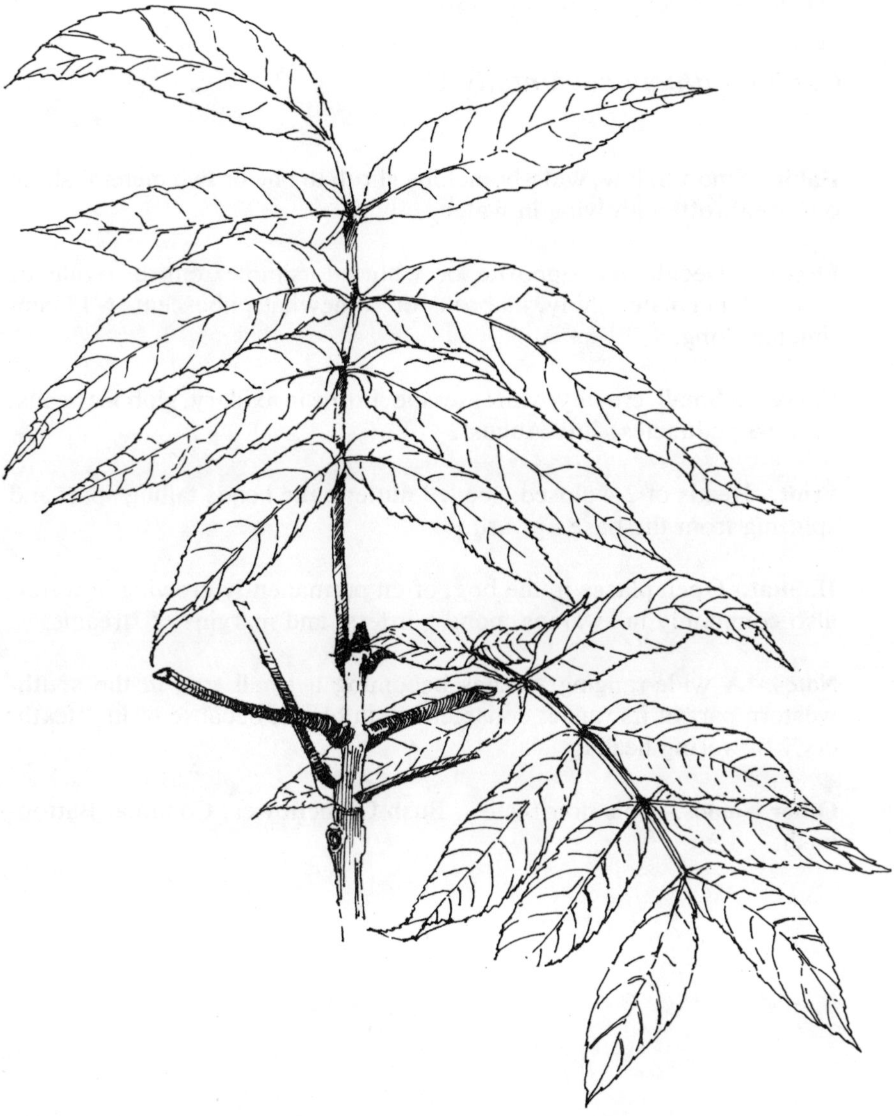

RUBIACEAE (Madder Family)
Buttonbush
Cephalanthus occidentalis L.

Habit.- Smooth, low, wide-branching shrub to one or two meters tall (in our area), often growing in water.

Leaves.- Deciduous, opposite or whorled, simple, entire, ovate or elliptic-lanceolate, shiny, glabrous or somewhat pubescent, 6-15 centimeters long.

Flowers.- Small, creamy-white, sessile, borne in axillary, globose heads, on long peduncles, July-August.

Fruit.- Heads of 2-4 closed, seeded nutlets, the heads falling apart and splitting from the base upward.

Habitat.- Open places in the bog, often permanently growing in water; also commonly in swamps, pond borders, and margins of streams.

Notes.- A wide-ranging species becoming a small tree in the southwestern part of its range; conspicuous in bloom because of its "feathery" flowering heads.

Other Names.- Bois noir (Que.), Bush Globeflower, Common Buttonbush.

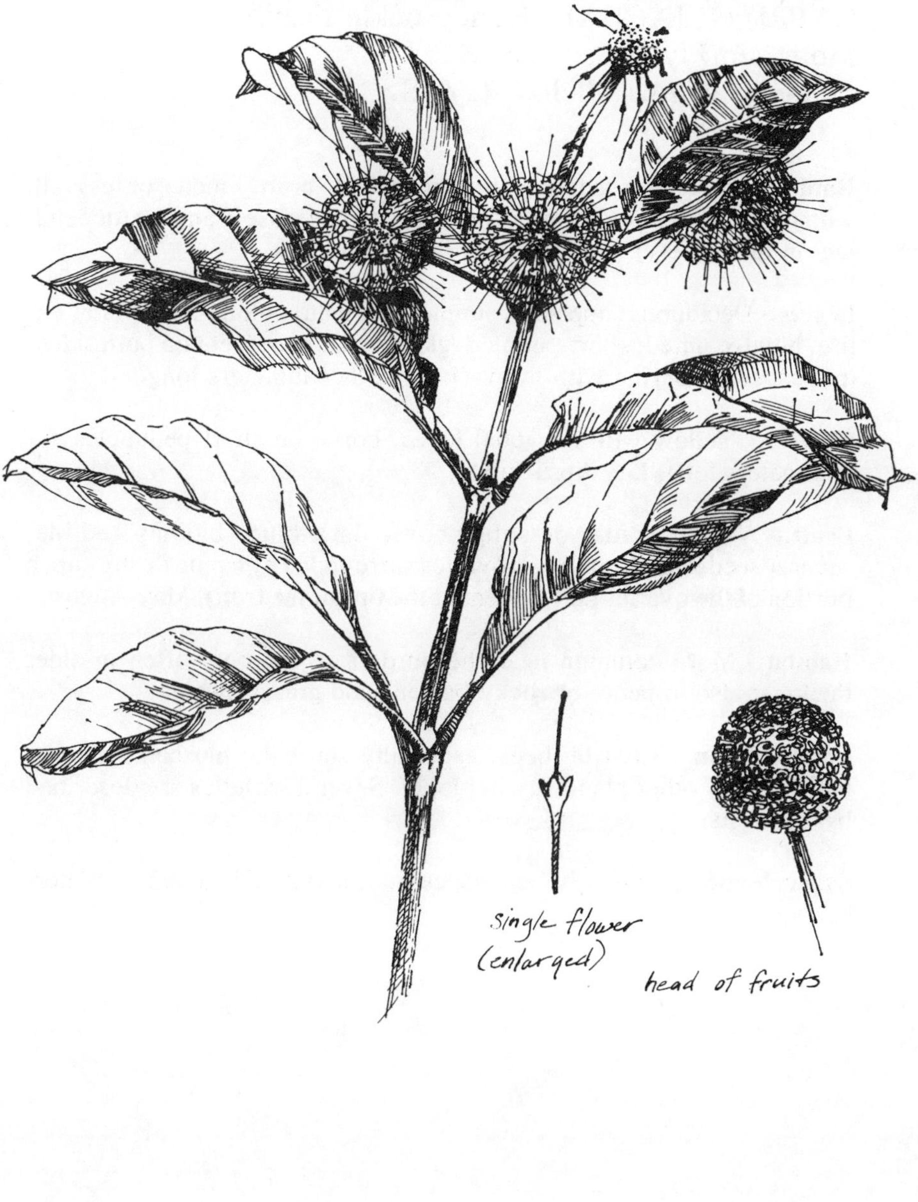

single flower
(enlarged)
head of fruits

CAPRIFOLIACEAE (Honeysuckle Family)
Mountain Fly-Honeysuckle
Lonicera villosa Michx. R. & S.

Habit.- Low, depressed ascending or upright shrub, 1 meter or less tall, with glabrous to more or less pubescent or tomentose branches (depending on the variety), and shreddy, tan-brown bark.

Leaves.- Deciduous, opposite, simple, oblong to oblong-lanceolate, entire, blunt-pointed, short-petioled, glabrous or pubescent on both sides, (pubescence varying with the variety), 2-8 centimeters long.

Flowers.- Yellow with subequal lobes, borne on short peduncles 1-7 millimeters long, late April-July.

Fruit.- An ellipsoidal or subglobose, dark blue, bloomy, edible, several-seeded berry (the two ovaries surrounded by a blue fleshy cup, a portion of the ovaries still evident at the tips of the fruit), May-August.

Habitat.- More common near the borders of the bog, often in alder thickets; also in peaty or rocky barrens and grassy swales.

Notes.- Often visited by bees, especially since the blossoms emerge before many other plants are in bloom. Several varieties are described by botanists.

Other Names.- Blue Fly-honeysuckle, Chèvrefeuille (Que.), Waterberry.

CAPRIFOLIACEAE (Honeysuckle Family)
Swamp Fly-Honeysuckle
Lonicera oblongifolia (Goldie) Hook.

Habit.- Erect shrub to 1.5 meters high, with minutely pubescent, ascending branchlets.

Leaves.- Deciduous, opposite, simple, entire, oblong to narrowly obovate, short-pubescent on both sides, bluish-green above and grayish green below, 3-8 centimeters long.

Flowers.- Corolla yellowish-white, deeply 2-lipped, borne on filiform peduncles, 1.5-4 centimeters long, late May-June.

Fruit.- An orange-yellow to deep red, subglobose berry, borne singly or united in pairs with few to many seeds, July, August.

Habitat.- Places in the bog, often near Northern White-cedar (*Thuja occidentalis* L.) where the area is underlain with limestone; also swampy thickets and wet woods.

Notes.- Occasional, especially northward and in calcareous areas.

Other Names.- Chèvrefeuille (Que.).

CAPRIFOLIACEAE (Honeysuckle Family)
Witherod
Viburnum cassinoides L.

Habit.- Upright shrub 1-4 meters tall with rather stiff branchlets and scurfy twigs.

Leaves.- Deciduous, opposite, simple, oblong or lanceolate to oval or ovate, rather firm and leathery, with a crenulate, dentate, or partially entire margin, 2.5-15 centimeters long and 1.5-6 centimeters broad.

Flowers.- Ill-scented, white or yellowish-white, June-early August.

Fruit.- Ellipsoidal to somewhat spherical drupe, changing from white through pink to blue-black, bloomy, 6-9 millimeters long; stone (pit) flat, elliptic-oblong, September-October.

Habitat.- Borders and open places in the bog, where it survives as a much dwarfed plant; also thickets, clearings, swamps, and borders of woods.

Notes.- Ripe pulp of fruit sweet; plant sometimes used around the home and for landscaping estates.

Other Names.- Alisier (Que.), Wild Raisin.

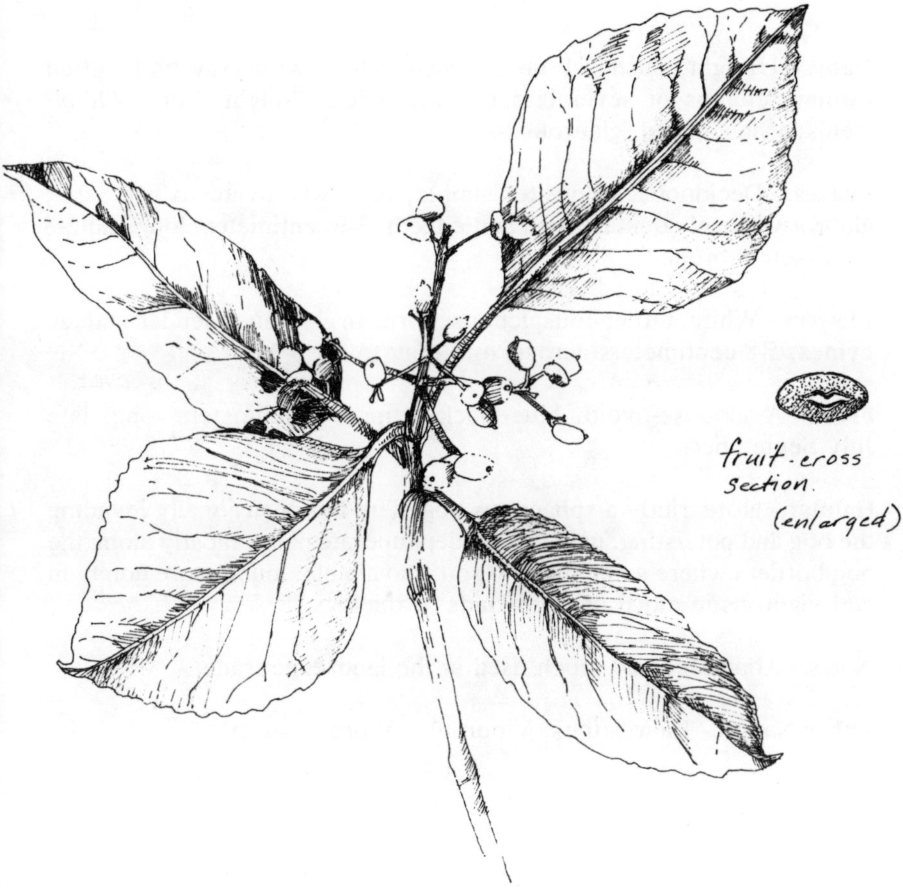
fruit-cross
section.
(enlarged)

CAPRIFOLIACEAE (Honeysuckle Family)
Arrow-wood
Viburnum recognitum Fern.

Habit.- Upright, bushy shrub 1-3 meters high with gray bark, often forming clumps of several to many slender, straight, "arrow-like" stems; twigs angled, glabrous.

Leaves.- Deciduous, opposite, simple, narrowly ovate to orbicular, glabrous, closely dentate with acute teeth, 3-9 centimeters long and 2-8 centimeters broad.

Flowers.- White, rather conspicuous, borne in glabrous, slender-stalked cymes, 5-8 centimeters across, mid-May to early July.

Fruit.- A globose-ovoid, blue-black drupe, 6 millimeters long, late July-September.

Habitat.- Not strictly a sphagnous bog plant but occasionally invading the bog and persisting, usually as a depauperate plant; mostly along the bog borders where some mineral soil is available; much more common and vigorous in moist or dry woods or thickets.

Notes.- Adaptable and much used in the landscape trade.

Other Names.- Laurestinus, Viburnum, Viorne (Que.).

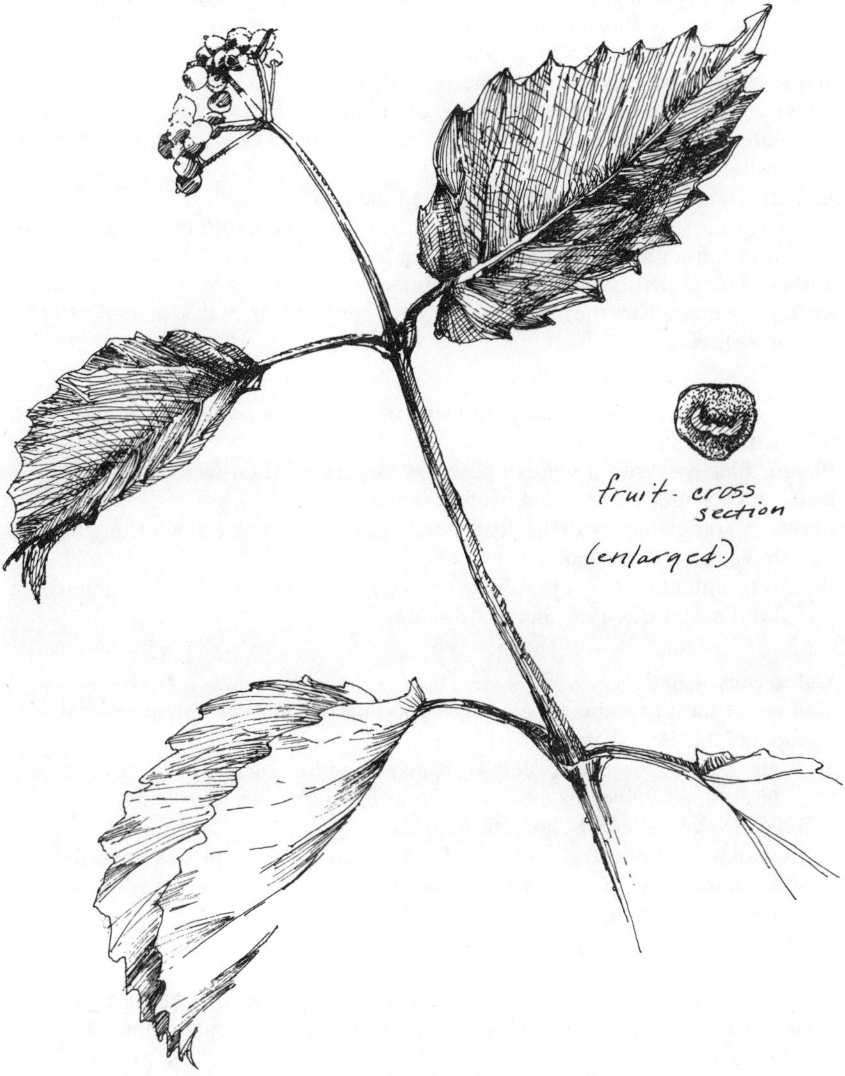
fruit - cross
section
(enlarged)

Glossary

Achene. A small, dry, hard, 1-celled, 1-seeded fruit which does not split open along definite lines at maturity.

Acuminate. Tapering to a long, sharp point.

Acute. Sharp-pointed, without tapering sides.

Adnate. United to an organ of a different kind.

Alternate. Said of leaves, buds, or leaf scars occurring singly at different points on the twig or branchlet.

Ament. Catkin, usually unisexual, often flexuous.

Angiosperm. Subdivision of seed-plants in which the ovule or ovules are enclosed in an ovary which becomes a fruit at maturity.

Anther. The pollen-bearing part of a stamen.

Axillary. Situated in the upper angle between bud or leaf scar and twig or branchlet.

Berry. A simple fleshy or pulpy fruit with usually many seeds. It contains no pit (stone) or core.

Bloom, Bloomy. With a whitish powdery covering, often waxy in nature.

Boss. A knob-like or rounded protuberance.

Bract. A somewhat modified leaf, occurring below and often belonging to a flower or inflorescence.

Bud. A rudimentary twig or growing point containing undeveloped vegetative or floral parts, often protected by scales.

Calcareous. Limy.

Callous. A hard protuberance or projection firmer than the surrounding tissue.

Campanulate. Bell-shaped.

Capsule. A dry, many-seeded fruit which splits open along several pre-determined valves or cracks at maturity to liberate the seeds.

Ciliate. Fringed along the margin with hairs.

Compound. Said of a leaf which is divided into separate blades or leaflets.

Conifer, Coniferous. Cone-bearing woody plant.

Connate. United to another similar or like structure.

Corolla. A collective term for all the petals.

Crenate. Bearing rounded, often glandular, teeth.

Cuneate. Wedge-shaped, narrowly triangular; a term used to describe leaf bases.

Cyme. A broad or flat flower cluster with the central or terminal flowers blooming earliest.

Deciduous. Falling from a plant as the result of a natural process called "abscission"; not evergreen.

Dehisce, Dehiscent, Dehiscing. Opening along certain pre-determined lines or cracks in order to liberate the seeds.

Dentate. Toothed, with the teeth directed outward.

Diaphragmed. A solid pith having regularly-spaced horizontal partitions.

Dioecious. Unisexual, with the two kinds of flowers on separate plants.

Drupe, Drupaceous. A simple fleshy fruit with usually one or few stony pits. (The pit is not the seed proper but is the hard, stony inner fruit wall which surrounds the usually one seed.)

Elliptic, Elliptical. Tapering uniformly toward both rounded ends, broadest in the middle.

Entire. Without teeth or lobes of any sort.

Erose. With a jagged margin, as if weathered or eroded.

Ferruginous. Rust-colored.

Follicle. A simple, dry fruit opening only along one side to discharge its seeds.

Fruit. The seed-bearing structure of a plant.

Glabrate. Becoming glabrous (free from hairs) with age.

Glabrous. Smooth, not hairy.

Gland, Glandular. A secreting surface or structure.

Glaucous. Covered with a bluish-white or bluish-gray substance, often waxy in appearance.

Globose. Spherical.

Gymnosperm. A woody plant bearing naked seeds, without an ovary.

Heterogeneous. Not uniform in kind.

Homogeneous. All alike or of one kind.

Imbricate. Overlapping.

Inflorescence. The flowering part of a plant, especially the mode of its arrangement.

Lanceolate. Lance-shaped, several times longer than wide, broadest below the middle and narrowing to the apex.

Linear. Long and narrow, with nearly parallel margins.

Mesophyte, Mesophytic. Plants inhabiting medium or favorable sites and conditions as to moisture and light.

Midrib. The central or main rib of a leaf.

Mucronate. Ending in a short and small, abrupt tip.

Node. The point on the twig which normally bears one or more leaves.

Nutlet. A small, hard, dry, one-seeded fruit; a diminutive nut.

Oblanceolate. Lance shaped with the broadest part toward the apex.

Oblong. Two or three times longer than broad and with nearly parallel sides.

Obovate. Having the outline of a hen's egg, with the broader end toward the tip.

Obsolescent. Becoming rudimentary or extinct.

Obtuse. Blunt or rounded at the end.

Opposite. Two at a node directly across the axis from each other.

Ovary. The part of the flower (base of the pistil) which contains the ovules (potential seeds).

Ovate. Having the outline of a hen's egg, with the broader end toward the base.

Panicle. A loose or branched inflorescence with stalked flowers.
Pedicel. The stalk or stem which supports a single flower.
Peltate. Shield-shaped and attached by the lower surface.
Petiole. Leaf stalk.
Pilose. Covered with soft hairs.
Pistil, Pistillate. The seed-bearing organ of a flower.
Pollen, Pollen grains. Spores or grains borne by the anther (swollen portion of the stamen), which produces the male gametes of the flower.
Polygamous. With perfect and unisexual flowers on the same or on different individuals of the same species.
Pome. A fleshy fruit, such as the apple; a core-fruit.
Prickle. A weak, slender, sharp outgrowth of the epidermis or bark.
Puberlous. Finely pubescent.
Puberulous, Puberulent. Minutely pubescent with short hairs.
Pubescent. Covered with fine, short, soft hairs.
Pulverulent. Appearing powdered with minute particles of dust.
Pyriform. Pear-shaped.

Rachis. The axis of an inflorescence or of a compound leaf.
Receptacle. The expanded or enlarged tip portion of the axis which bears the organs of a flower or collection of flowers.
Revolute. With margins rolled toward the lower side.
Rhizome. A prostrate or subterranean stem, usually rooting at the nodes.
Roseate. Rose-colored or rose-tinted.
Rugose. Wrinkled.
Rugulose. Shallowly or obscurely wrinkled.

Seed. The ripened ovule, consisting of embryo and seed coats, with or without endosperm (accumulated food).
Serotinous. Produced late in the season.
Serrate. Having sharp, forward-pointing teeth.
Sessile. Without a stalk.
Setose. Beset with bristles.
Shrub. A woody perennial, smaller than a tree, usually with several stems.
Simple. Said of a leaf with only one blade; not compounded into separate leaflets.
Sordid. Impure white, appearing dirty.
Stamen, Staminate. The pollen-bearing organ of a flower.
Stipule. An appendage at the base of the petiole or leaf or on each side of its insertion.
Stolon, Stoloniferous. A runner or any basal branch that is inclined to root.
Strigose. Beset with appressed, sharp, straight, and stiff hairs.
Strobile, Strobilus. An inflorescence composed of overlapping bracts or scales.
Style. The usually attenuated portion of the pistil connecting the stigma with the ovary.
Subglobose. Somewhat or slightly globose or spherical.
Suborbicular. Somewhat orbicular or circular.
Subshrub. A barely or only slightly woody plant or shrub.
Suture. A line or place of splitting.

Terete. Circular in cross-section.
Ternate. In threes.
Tetraploid. Having four times the basic (or twice the diploid) number of chromosomes.
Tomentose. Covered with dense, matted, woolly hairs.
Tomentulose. Minutely woolly.
Tomentum. A dense covering of matted hairs.
Tree. A woody plant potentially exceeding twenty feet in height and usually with a single trunk and a definite crown.

Urceolate. Urn-shaped.

Vein. Thread of fibrovascular tissue in a leaf.
Viscid. Glutinous, sticky.

Whorled. Occurring three or more at a node.

INDEX TO FAMILIES AND SPECIES
(Including both technical and common names)